Merrow Technical Library
Practical Science

SELECTIVE ION SENSITIVE ELECTRODES

Merrow Technical Library

General Editor: J. Gordon Cook, B.SC., PH.D., F.R.I.C.

Practical Science

SELECTIVE ION SENSITIVE ELECTRODES

G. J. Moody, B.Sc., Ph.D.
and
J. D. R. Thomas, MSc., F.R.I.C.

University of Wales Institute of Science and Technology, Cardiff, Wales

MERROW

Merrow Publishing Co. Ltd.
276 Hempstead Road
Watford   Herts   England

ISBN 0 900 54135 0

Printed in Great Britain at the Pitman Press, Bath

FOREWORD

The development of selective ion sensitive electrodes designed to respond to a particular ion in solution has been heralded as a "whole new technology". About twenty cations and anions can now be individually determined as easily as hydrogen ions by the classical pH glass electrode, from the magnitude of the electrical potential relative to suitable reference electrodes. The various selective ion electrodes comprise cation glass, silicone rubber, solid state and liquid ion exchanger types.

All branches of science have benefited from the commercial availability, versatility and ease of operation of selective ion sensitive electrodes. Indeed, their impact on solution chemistry has been likened to that of the laser on optical physics.[1] Routine analysis in medical research includes saliva, cerebrospinal fluid, serum, urine, sweat, bone enamel and general dentistry. The electrodes are also useful for quality and general process control of pharmaceuticals (drugs, soaps and toothpastes) and foodstuffs (dairy, cereals, grain, sugar, beverages and canned produce), and for mineral assay of rocks, clays, ore refining processes, industrial and mining wastes.

Other important industrial applications involve plating baths and metal finishings, rinse tanks, etching baths for printed board manufacture, paints, pigments, algaecides, fungicides, pesticides, fertilizers, ceramics, glass, photography, explosives, liquid propellants, stack gas effluents, tobacco, plastics, resins, greases, oils, lubricants and paper pulp liquors. Ecological factors, such as analyses of streams, rivers, estuaries, lakes, sewage, mud, soils, as well as water softening, and scale prevention in boilers and cooling plants, have been investigated. At the purely academic level, activity coefficients, complexation reactions, stability constants, and potentiometric titrations have been studied.

The fact that some of the performances listed for these electrodes are based on what the manufacturers "feel" they will be best suited for, should be borne in mind by any prospective user. This monograph covers both theoretical and practical aspects of selective ion sensitive electrodes. No complete historical coverage or exhaustive inclusion of journal and manufacturing listings is intended.

G.J.M.

July 1970   J.D.R.T.

CONTENTS

1

MEMBRANE POTENTIAL

The utility of membrane electrodes depends upon the determination of membrane potentials. Such potentials cannot be measured directly, but can be deduced from the e.m.f.s of complete electrochemical cells. Thus, the membrane potential in the cell

Electrode A|Solution 1|Membrane|Solution 2|Electrode B

Electrode Potential   Membrane   Electrode Potential
                       Potential

may be regarded as the electric potential difference between the two bulk solutions 1 and 2, and includes both the diffusion potential within the membrane and the phase boundary potentials.

In order to obtain the membrane potential from the cell e.m.f., assumptions have to be made about the electrode potentials, and a straightforward procedure is the use in e.m.f. measurements of calomel electrodes (or other reversible reference electrodes) (Fig. 1) for electrodes A and B.

$Hg|Hg_2Cl_2|KCl$ sat.|Solution 1|Membrane|Solution 2|KCl sat.|Hg_2Cl_2|Hg

For simplicity, the cell may be likened to a concentration cell without transport where the two solutions consist of the same electrolyte but in different concentrations. The membrane potential in such a system may, therefore, be regarded as the concentration potential. It should then be possible to relate the membrane potential (E_M) to ion activity just as in concentration cells:

1

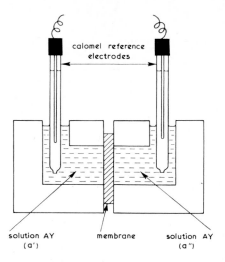

Fig. 1. *Typical cell arrangement for measuring EMF of membrane electrode.*

$$E_M = \frac{2 \cdot 303\,RT}{z_A F} \left(\log \frac{a_A''}{a_A'} \right) \tag{1}$$

where a_A is the activity of the ion A and z_A is its valence.

However, eqn(1) is an over-simplification, and in addition to the Donnan phase boundary potential, should include a term to allow for diffusion phenomena arising from the fact that the membrane is more permeable to counter ion A than to the co-ion Y:

$$E_M = \frac{2 \cdot 303\,RT}{z_A F} \left[\log \frac{a_A''}{a_A'} - (z_Y - z_A) \int_,^{''} t_Y\, d \log a^{\pm} \right] \tag{2}$$

where t_Y is the transport number of the co-ion in the membrane and z_Y its valence; $a^{\pm}$ is the mean ionic activity of the electrolyte.[2] The integral may be evaluated graphically.

Expressed in another way, the first term of eqn(2) gives the thermodynamic limiting value of the concentration potential, while the second term gives the deviation due to the co-ion flux. With a membrane ideally perm-selective towards the counter ion, the second term vanishes $(t_y = 0)$ and eqn(2) reduces to the Nernst form, i.e. eqn(1).

With the usual ion exchange membranes, eqn(1) holds reasonably well between about 10^{-4} and 10^{-1} M. Deviations at higher solution activities are caused by co-ion transference, and at lower activities by competing hydrogen or hydroxyl ions which stem from dissociation of water.

Within the valid range of the limiting eqn(1), an ion exchange membrane can be used for determining ionic activities. The measurement is made in a cell (Fig. 1) with a membrane which is ideally permselective for the ion A. One solution compartment of the cell is filled with a standard solution of known activity a'_A. The unknown activity, a''_A, in the solution on the other membrane side can be calculated from the membrane potential using eqn(1).

The half-cell

$$\text{Hg}|\text{Hg}_2\text{Cl}_2|\text{KCl sat.}|\text{Solution } A\,Y|\text{Membrane}\|$$

thus acts as an electrode which is reversible with respect to the ion A and is known as a *membrane electrode*.

Membrane electrodes have the great advantage that they can be built for almost any ion. Unfortunately, the difficulty of distinguishing adequately between different ions, even of equal sign, is a serious drawback, and to date only about twenty cations or anions can actually be evaluated. Nevertheless, as indicated above, the developments even with this limited number of ions are of immense practical importance and the success of the glass pH electrode is illustrative of the tremendous future prospects of these selective ion-sensing devices.

3

2

PRINCIPLES OF SELECTIVITY ASSESSMENT

Functional Potential

The functional potential of any selective ion-sensitive membrane electrode depends on a number of factors including potential-activity responses, selectivity in the presence of various interferants, operative pH range, response times, temperature and operative life. With few exceptions, the electrodes follow a Nernstian potential-activity pattern within working range as given by the relation

$$E = E° + \frac{2\cdot303\, RT}{zF} \log a_{cation} \tag{3}$$

for cationic responsive electrodes (Fig. 2) and

$$E = E° - \frac{2\cdot303\, RT}{zF} \log a_{anion} \tag{4}$$

for anionic responsive (Fig. 3) electrodes.

If concentration assessment is to depend on the convenient linear relationship between response and log activity, then knowledge of the activity/concentration relationships as given by the Debye-Hückel relationship is demanded, otherwise concentration has to be assessed from a carefully plotted direct response/concentration calibration curve known as a *working curve*. The working curve is not a straight line (concentration curves of Figs. 2 and 3) and does not have a Nernst slope. Furthermore, it will be in error if the level of the ions in the sample solution is high; this is because the activity (which is related to e.m.f. response) depends on the total ionic strength.

4

Activity and Activity Coefficient

The activity coefficient, f, of an electrolyte is related to the activity, a, since it is equal to the latter divided by the measurable concentration of the substance.

Hitherto, the activity of an electrolyte was regarded as a purely thermodynamic quantity which can be evaluated from the observable properties of the solution. However,

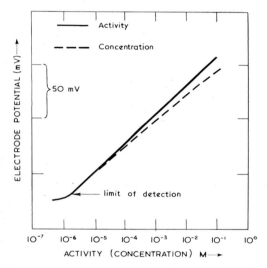

Fig. 2. Typical activity (concentration) calibration graph for Coleman solid state copper electrode.

within the present context interest centres on concentration, so that, unless resort is made to carefully plotted e.m.f. response/concentration working curves, users of selective ion-sensitive electrodes have to be familiar with the relationship between activities and concentration if they are to make use of the convenient Nernst slopes characteristic of e.m.f. response/log a plots. In this connection, it is noteworthy

5

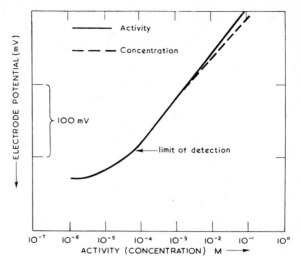

Fig. 3. Typical activity (concentration) calibration graph for Philips solid state cyanide electrode.

that the theory of Debye and Hückel[3] made it possible to calculate the mean ionic activity coefficient, f, for a dilute solution of an electrolyte without recourse to experiment.

By a theoretical argument based on electrostatics and statistical mechanics, Debye and Hückel derived the following approximate equation for determining ionic activity coefficients:

$$\log f = -Az^2\sqrt{\mu} \qquad (5)$$

where f is the activity coefficient of the ion, z is its charge, μ the ionic strength of the solution and A is a constant which depends on temperature and solvent ($A = 0\cdot511$ for water at 25°C).

The Debye-Hückel equation is a limiting law, correct in the limit of zero ionic strength, which predicts that the activity coefficient of an ion depends only on its charge and

6

the ionic strength of the environment; that the identity of the other ions in the solution is unimportant beyond the specification of their charge; and that f will decrease uniformly as μ increases (but see Figs. 4 and 14). All these predictions are true at extremely low ionic strength.

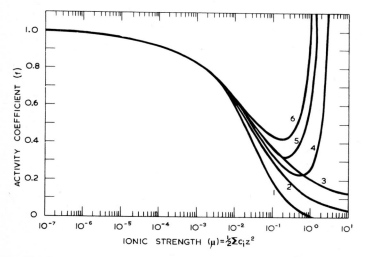

Fig. 4. Computer simulations of various activity coefficient-ionic strength equations for divalent ions. (Each plot represents 800μ-f pair values.)

At ordinary concentrations, for example at 10^{-1} M, the behaviour of real electrolytes deviates from the relatively simple predictions of the Debye-Hückel limiting law. Several more or less empirical extensions have been proposed; they fit the data with varying success at moderate values of the ionic strength, μ. It is always possible to fit any set of empirical data with an equation which contains a large number of adjustable empirical parameters; such equations are neither aesthetically appealing nor generally satisfactory, since they require the measurement and tabulation of otherwise meaningless constants for each ion studied. Because many situations confront an investigator with ions which have not been so studied, it would be best to have an equation which is applicable—at least approximately—to any ion. Of those which have been proposed, the simplest are based on the following modification:[4]

$$\log f = -Az^2 \left(\frac{\mu^{\frac{1}{2}}}{1 + \mu^{\frac{1}{2}}} - 0.2\mu \right) \tag{6}$$

Plots based on various equations are illustrated in Fig. 4, the ionic strength, μ, being computed by taking half the sum of the terms obtained by multiplying the concentration of every ion, c, present in solution by the square of its valence, z:

$$\mu = \tfrac{1}{2}\Sigma cz^2 \tag{7}$$

A calculation of ionic strength using eqn(7) is shown in Table 1.

TABLE 1

Calculation of Ionic Strength

Solution Composition (mol litre^{-1})	Ions	z^2	C	cz^2
0·05 NaCl + 0·20 CaCl$_2$ + 0·15 NaNO$_3$	Na$^+$	1	0·05 + 0·15	0·20
	Ca^{2+}	4	0·20	0·80
	Cl$^-$	1	0·05 + 0·40	0·45
	NO$_3^-$	1	0·15	0·15
			Σcz^2	1·60

Therefore, total ionic strength, $\mu = \tfrac{1}{2}\Sigma cz^2 = \dfrac{1\cdot60}{2} = 0\cdot80$

8

Selectivity

If the full scope of selective ion-sensitive electrodes is to be usefully realized, their selectivity in the presence of various interferants demands proper and reliable assessment. Unfortunately, there is little agreement in the literature regarding optimal methods for determining or quoting selectivities.

A selectivity constant indicates the extent to which a foreign ion N^{n+} interferes with the response of an electrode to its primary ion M^{z+}. In general, for electrodes responding to cations, the constant K_{MN} is defined by

$$E = E^\circ + \frac{2\cdot303\ RT}{zF} \log [a_{M^{z+}} + K_{MN}\,(a_{N^{n+}})^{z/n}] \qquad (8)$$

Except for the matter of sign, the relation for anions is similar and has been discussed by Srinivasan and Rechnitz[5] for the case where both the primary and interferant anions are univalent.

For all values of $K_{MN} < 1$, the better the electrode, at least with respect to selectivity favouring the species M^{z+}. Thus, the selectivity constant $K_{NO_3CN} = 2 \times 10^{-2}$ for the Orion 92–07 nitrate electrode, means that it is fifty times more sensitive to nitrate than to cyanide ions. On the other hand, $K_{NO_3ClO_4}$ is 10^3 and the title nitrate electrode is underserved in this context since it is now 1000 times more sensitive to perchlorate than to nitrate ions.

Unqualified use of the term *selectivity constant* can be misleading in several ways. Firstly, the value for any two ions is activity dependent. Secondly, any quoted value depends on whether potentials are taken in mixed or separate solutions containing the two ions, as well as on the equation employed in subsequent calculations. An additional problem with selectivities is that they are occasionally given as reciprocals (K_{NM}) or, worse, listed simply as K and/or without a mention of the particular evaluation method.

The various methods for evaluating selectivities, based on

9

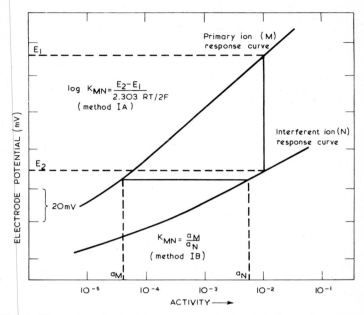

Fig. 5. Illustration of selectivity constant (K_{MN}) calculations using separate solution method with divalent ions.

potential measurements in separate and mixed solutions, will now be discussed.

Method I. Separate Solution Method
The e.m.f. response to various activities of a primary cation, say M^{2+}, are plotted as in Fig. 5. A similar plot is made for the interferant cation, N^{2+}. Selectivities can then be evaluated in two ways.

Method IA. The e.m.f. response, E_1, to the primary divalent cation is

$$E_1 = E^\circ + \frac{2 \cdot 303\, RT}{2F} \log a_{M^{2+}} \tag{9}$$

10

Ranges of Selectivity Constants K_{CaN} for Orion 92–20 and PVC Calcium Electrodes Measured in Separate and Mixed Solutions (6)

Interferant Ion N^{n+}	Separate Solution Method*				Mixed Solution Method IIA (Method IIB for H^+)					Manufacturer's Values (Orion 92–20 Electrode)
	Orion 92–20 Electrode		PVC Electrode		Molar Range of Interferant N^{n+}	From Extrapolated Intercept †		From 18/z		
	Method IA	Method IB	Method IA	Method IB		Orion 92–20 Electrode	PVC Electrode	Orion 92–20 Electrode	PVC Electrode	
Mg^{2+}	0·01 → 0·29	0·007 → 0·04	0·005 → 0·13	0·0045 → 0·0097	4·3 × 10⁻⁵ → 5 × 10⁻³	0·055 → 0·025	0·222 → 0·024	0·036 → 0·016	0·14 → 0·025	{ 0·014, 0·01
Ba^{2+}	0·0059 → 0·18	0·0036 → 0·027	0·003 → 0·090	0·002 → 0·0074	5 × 10⁻³ → 5 × 10⁻²	0·033 → 0·011	0·013 → 0·005	0·034 → 0·011	0·0072 → 0·004	0·01
Zn^{2+}	0·011 → 0·65	0·007 → 0·32	0·0098 → 0·90	0·0056 → 0·5	5 × 10⁻⁴ → 5 × 10⁻³	0·081‡	0·065‡	0·77	0·06	3·2
Na^+	0·0058 → 0·45	0·005 → 0·006	0·0052 → 0·27	0·0054 → 0·0061	10°	5·3 × 10⁻⁴	2·1 × 10⁻⁴	1·1 × 10⁻⁴	5·8 × 10⁻⁵	{ 1·6 × 10⁻³, 3 × 10³
K^+	0·002 → 0·37	0·0015 → 0·0016	0·0026 → 0·39	0·0027 → 0·0055	10°	6·6 × 10⁻⁵	2·2 × 10⁻⁵	10⁻⁴ → 3 × 10⁻⁵		10⁻⁴
H^+	330 → 590	590	29 → 27	26	pH2 → pH9	1·3 × 10⁴ → 2 × 10³	40 → 25	=	=	{ 10⁵, 10⁷

* Activity range of N^{2+}, $1.9 \times 10^{-2} \to 9.7 \times 10^{-6}$ M and of N^+, $10^{-1} \to 10^{-2}$ M.
† Using same equation as used in Method IB.
‡ For the higher concentration only.
|| Dips prevent the use of this method.

The potential response, E, to a divalent cation mixture, $M^{2+} - N^{2+}$, is given by:

$$E = E^\circ + 2{\cdot}303 \frac{RT}{2F} \log [a_{M^{2+}} + K_{MN} a_{N^{2+}}] \qquad (10)$$

The potential response, E_2, to the N^{2+} cation can be obtained[5] from eqn(10) by making $a_{M^{2+}} = 0$:

$$E_2 = E^\circ + 2{\cdot}303 \frac{RT}{2F} \log K_{MN} a_{N^{2+}} \qquad (11)$$

Subtracting eqn(9) from eqn(11) gives:

$$E_2 - E_1 = 2{\cdot}303 \frac{RT}{2F} [\log K_{MN} + \log a_{N^{2+}} - \log a_{M^{2+}}] \qquad (12)$$

When the primary and interferant activities are equal, then eqn(12) simplifies to:

$$\frac{E_2 - E_1}{2{\cdot}303 RT/2F} = \log K_{MN} \qquad (13)$$

A large number of K_{MN} values can then be easily taken from the differences in potentials $E_2 - E_1$ at which the primary and interferant cations are at the same activities (Fig. 5).

Method IB. Essentially, this concerns an alternative treatment of the data in Fig. 5. When $E_1 = E_2$, eqns(9) and (11) can be combined to give:

$$\log a_{M^{2+}} = \log K_{MN} a_{N^{2+}} \qquad (14)$$

that is,

$$a_{M^{2+}} = K_{MN} a_{N^{2+}} \qquad (15)$$

When the ions have different charges, eqn(15) needs modification. For example, when the interferant has a charge $n+$, eqn(15) becomes

$$a_{M^{2+}} = K_{MN} (a_{N^{n+}})^{2/n} \qquad (16)$$

12

Thus, the selectivity is equal to the ratio of the activities which realize identical potentials in their respective separate solutions. Again, more than one K_{MN} value is possible, depending on the solution conditions (Fig. 5).*

Neither method is realistic since separate solutions are used. However, the selectivities so determined[5] for several Orion electrodes compare quite favourably with the values obtained in mixed solutions and, in any case, all methods show selectivity to be concentration/activity dependent.

* Selectivities are also quoted in a reciprocal fashion, that is, as the ratio of the interferant and primary ion activities which separately realize the same potential responses.

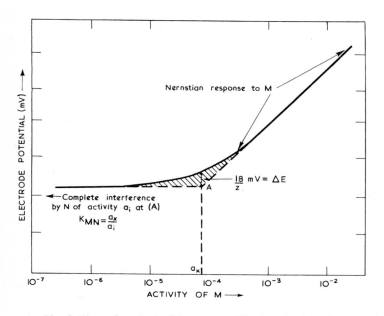

Fig. 6. *Illustration of selectivity constant (K_{MN}) evaluation using fixed initial amount of interferant.*

13

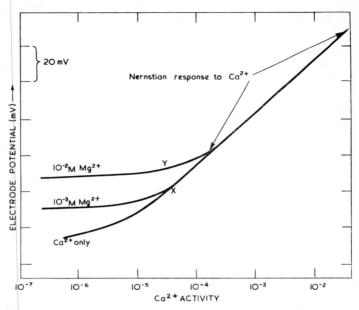

Fig. 7. Influence of magnesium on response characteristics of calcium-sensitive PVC electrode.

Method II. Mixed Solution Method

Method IIA. A very common procedure is to measure potentials in solutions containing a fixed initial amount of interferant, say N^{2+}, varying the activities of that ion, say M^{2+}, for which the electrode is designed. Fig. 6 illustrates the general idealized pattern. As the activity of the primary cation falls, there is gradual onset of interference (shaded zone) until eventually complete interference sets in along the horizontal plateau, when the potential is constant. The intercept of the extrapolated Nernstian response line with that of the horizontal total interference defines[7] a particular intercept activity for the primary cation a_x (Fig. 6). The selectivity constant is then readily calculated using eqn(15)

14

as $a_x = Ka_i$, where a_i is the constant background interference activity.[7]

Under conditions of high interference, the potential measurements in the plateau region are subject to drift and irreproducibility. In such cases, the required a_x value is better located by taking the point on the experimental potential curve where ΔE between that curve and the extrapolated Nernstian line is exactly $18/z$ mV (Fig. 6, Table 2). The activity at this point will be the required a_x value as before.[7]

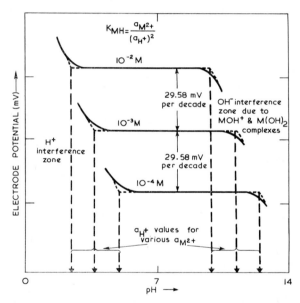

Fig. 8. A typical potential response pattern of M^{2+} responsive electrode in a solution of its pure M^{2+} salts at varying pH.

Many of the published selectivity values for glass and liquid ion exchange electrodes seem to have been evaluated by this simple mixed solution method.

15

Fig. 7 stresses the care needed in any interpretation of any one published selectivity as it will vary with the relative calcium-interferant activities. Thus, the value $K_{CaMg} = 0.024$ obtained at a high magnesium interference level ([Mg^{2+}] = $5 \times 10^{-3}M$) suggests a better selectivity than $K_{CaMg} = 0.222$ at a lower magnesium level ([Mg^{2+}] = $4.3 \times 10^{-5}M$), whereas the pvc electrode (p. 113) actually gives the most extensive Nernstian response at lower magnesium levels. (Fig. 7).

Method IIB. The technique employed in an alternative method is really the reverse of Method IIA. Thus, the interferant N^{2+} is varied against a constant level of primary ions M^{2+} (Fig. 8).

One of the most important interferences for an electrode is pH, and this method is extensively used for H^+

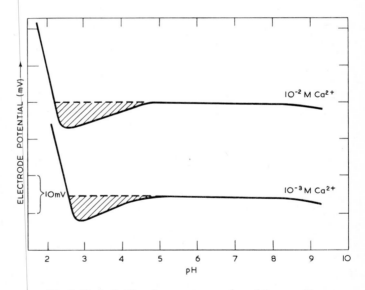

Fig. 9. Typical pH-emf response curve for calcium sensitive electrode containing Orion liquid ion exchanger.

16

interference (Fig. 8). However, K_{MH} values are also available and are calculated from the pH-potential graph with eqn(15) except that the interference activity, a_{H^+} is found from the intercept.

The effect of H^+ interference on the pvc and Orion 92-20 calcium electrodes has been made by following simultaneous potential and pH responses against a constant calcium chloride activity level after adding successive small increments of molar hydrochloric acid (Fig. 9). The characteristic dips (shaded) for both these calcium electrodes are unlike pH interference plots for other electrodes. The dips have been attributed to impurities, since claimed to have been removed, from some Orion liquid ion exchanger batches.[8]

A general feature of many such plots is the tailing at high pH levels, due in some cases to a drop in the cation activity following complexation to MOH^+ and $M(OH)_2$ species. Presumably, K_{MOH} values could be found from these tailings in a manner analogous to hydrogen selectivities.

Method IIC. Neither of the previous methods is considered by Srinivasan and Rechnitz[5] for their treatment of mixed solution potential data using Orion anion liquid ion exchange electrodes. Instead, two rather complicated equations have been devised, for high and low selectivities respectively.

The potential, E_1, is measured in the pure solution of the primary univalent anion M^- of activity a_{M^-}, when:

$$E_1 = E^\circ - 2 \cdot 303 \frac{RT}{F} \log a_{M^-} \qquad (17)$$

Known quantities of univalent interferant, N^-, are successively introduced so as to give a series of potential values for the mixed solutions. The potential, E_2, in any one of the measurement series is related to the new primary activity a'_{M^-} and interferant activity, a'_{N^-}, by:

$$E_2 = E^\circ - 2 \cdot 303 \frac{RT}{F} \log [a'_{M^-} + K_{MN} a'_{N^-}] \qquad (18)$$

17

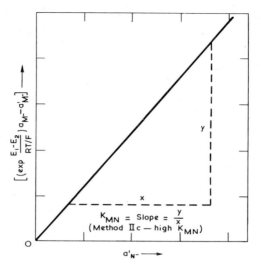

Fig. 10. Illustrative selectivity plot for N^- interferant with M^- ion-sensitive electrode.

Eqns(17) and (18) can be easily combined to give:

$$\left[\exp^{\frac{E_1 - E_2}{RT/F}} \right] a_{M^-} - a'_{M^-} = K_{MN} a'_{N^-} \qquad (19)$$

Plots of the left-hand side of eqn(19) against interferant activities are linear with slopes equal to the selectivity constant K_{MN} (Fig. 10).

However, where K_{MN} is small, the potential cannot be expected to be sensitive to small increases in the activity of interferant. In such cases, the procedure adopted[5] is to maintain a fairly high activity of interferant anion, and measure a series of potentials, E_2, after additions of primary anion. A slightly more complicated double exponential-term equation is necessary in these cases.[5]

18

$$a'_{M^-} - \left[\exp^{\frac{E_1 - E_2}{RT/F}}\right] a_{M^-} = K_{MN}$$
$$\left[\left(\exp^{\frac{E_1 - E_2}{RT/F}}\right) a_{N^-} - a'_{N^-}\right] \quad (20)$$

Plots of the complete left-hand side of eqn(20) against the square bracket function on the right-hand side also give good linear plots of slope equal to K_{MN}. The selectivities based on eqns(19) and (20) compare quite favourably with the two separate solution methods based on eqns(13) and (15) (Table 3).

TABLE 3

Selectivities K_{MN}, of Three 92-Series Orion Anion
Electrodes (Reference 5)

| Orion Electrode | Interferant N | Separate Solutions | | Mixed Solutions | Orion values* |
		Method IA (eqn(13))	Method IB (eqn(15))	Method IIC (eqn(19))	
92-07 NO$_3$	ClO$_3^-$	1·23–1·41	1·22–1·45	1·14–1·23	2
	Br$^-$	0·15–0·24	0·13–0·18	0·09–0·13	0·9
	I$^-$	8·9–22·2	2·7–16·3	3·2–15·7	20
92-17 Cl$^-$	Br$^-$	1·72–2·88	2·04–3·73	2·58–3·42	1·6
	NO$_3^-$	1·68–4·09	2·20–5·89	2·62–4·67	4·2
	I$^-$	4·4–17·1	4·8–26·7	6·0–23·9	17
92-81 ClO$_4$	I$^-$	0·015–0·071	0·012–0·030	0·020–0·023†	0·012

* Presumably based on mixed solution method IIA.
† Using equation 20.

The mixed solution selectivities[5] are on a par with the separate solution values,[5] and, except for bromide, not too different from the independent Orion values (Table 3). There seems little point, therefore, in using the arithmetically cumbersome mixed solution method of Srinivisan and Rechnitz[5] for evaluating selectivities. Furthermore, their plots offer no "at a glance idea" of selectivity as provided by the previous mixed solution plot (Fig. 7). It also appears that some prior knowledge of actual selectivity is necessary before the proper activities of primary and interferant ions can be chosen for the mixed solution potential measurements.

The shorter alternative treatments of mixed solution data (Methods IIA and IIB) are to be preferred in that no prior knowledge of selectivities is required. Separate solution methods also give very similar selectivities to those in mixed solution, at least for four of the Orion 92-series electrodes.[5,6] All the methods discussed will give a selectivity range, although there are numerous literature and even trade listings quoting a single numerical value for selectivity constants. Several manufacturers present actual mixed solution plots analogous to Figs. 7 and 9, as well as single value selectivity lists, in their trade literature. Unfortunately, but understandably, this ideal graphical presentation is not so common in journals. It is strongly recommended that users of these ion sensitive electrodes set up their own "at a glance" plots for any investigations involving interference parameters.

The selectivity constants for liquid membrane electrodes listed by manufacturers, as well as those from independent studies, are collected in Tables 4 and 5. These are single selectivity values, that is, for a primary ion and one interfering ion. There is little information on selectivities in the presence of more than one interferant species. In general, cations do not interfere with anion selective electrodes, and conversely, anions do not interfere with cation selective electrodes. Notable exceptions are iodide and perchlorate which at $> 10^{-3}$ M adversely affect the Orion 92-20 calcium electrode, but not the 92-32 divalent electrode.

Despite the variance of selectivity constants, they are extremely useful in pre-judging a particular analytical issue. The severity of interference depends on the relative activity levels of primary and interfering ions, and on the selectivity constant. For example, the percentage error incurred by a nitrate electrode in the presence of nitrate and any interferng anion, N^{n-}, can be readily found from:

20

Selectivity Constants* of Cation Liquid Ion Exchange Membrane Electrodes

Interfering Cation	Manufacturers' Electrodes								
	Orion Cu^{2+} 92–29	Orion Pb^{2+} 92–82	Orion Ca^{2+} 92–20	Orion Ca^{2+} 92–20†‡	Beckman Ca^{2+} 39608‖	Beckman Ca^{2+} 39608¶	Corning Ca^{2+} 476041**	Orion $Ca^{2+}-Mg^{2+}$ 92–32	Beckman $Ca^{2+}-Mg^{2+}$ 39614
Zn^{2+}	10^{-3}; 0·03	3×10^{-3}	3·2					3·5	> 1·0
Fe^{2+}	1·0; 140	0·08	0·80					3·5	
Pb^{2+}			0·63						
Mg^{2+}	<10^{-4}; 10^{-3}, 10^{-4};	8×10^{-3}	0·01; 0·014	0·04	0·12	0·34	0·01	0·94	0·95
Ba^{2+}	2×10^{-4}	0·01	0·01	0·04	0·079	0·90	0·01		0·8
Sr^{2+}	10^{-4}; 2×10^{-4}		0·017; 0·02	0·07	0·098		0·01	0·54	
Ni^{2+}	5×10^{-3}; 0·01	7×10^{-3}	0·08				0·01	1·35	
Cd^{2+}				0·03					
Ca^{2+}	5×10^{-4}; 2×10^{-3}	5×10^{-3}							
Cu^{2+}		2·6	0·27					3·1	
Na^+	<10^{-3}; 5×10^{-4}	$1·6 \times 10^{-3}$; 3×10^{-3}		10^{-3}	0·015	0·029	10^{-3}	0·01; 0·015	0·013
K^+	<10^{-3}; 5×10^{-4}	10^{-4}		10^{-3}		0·034	10^{-3}	<0·015	0·013
NH_4^+		10^{-4}							
H^+	10	10^5; 10^7			72	$1·5 \times 10^4$			

* Manufacturers' data unless otherwise quoted.
† Reference 9.
‡ See also data in Table 2 and reference 6.
‖ Reference 10.
¶ Reference 11.
** The selectivity of the analogous Corning divalent electrode (No. 476235) is simply listed as: $Ba^{2+} = Sr^{2+} > Ni^{2+} = Ca^{2+} = Mg^{2+} > Zn^{2+} = Cu^{2+} = Ca^{2+} : Ca^{2+} > (Na^+, K^+)$.

TABLE 5

Selectivity Constants* of Anion Liquid Ion Exchange Membrane Electrodes

Interfering Anion	Manufacturers' Electrodes						
	Orion BF_4^- 92–05	Orion BF_4^- 92–05†	Beckman BF_4^- 39620	Corning NO_3^- 476134‡	Orion NO_3^- 92–07	Orion NO_3^- 92–07¶	Beckman NO_3^- 39618
F^-	10^{-3}	9×10^{-5}	2×10^{-4}		$\begin{cases}6 \times 10^{-5}\\9 \times 10^{-4}\end{cases}$	5×10^{-4}	$6 \cdot 6 \times 10^{-3}$
Cl^-	10^{-3}		5×10^{-4}	4×10^{-3}	$\begin{cases}6 \times 10^{-3}\\4 \times 10^{-3}\end{cases}$	8×10^{-3}	$0 \cdot 02$
Br^-	$0 \cdot 04$		$0 \cdot 02$	$0 \cdot 011$	$\begin{cases}0 \cdot 9 – 0 \cdot 13\\0 \cdot 09 – 0 \cdot 24\|\|\end{cases}$		$0 \cdot 28$
I^-	20		$0 \cdot 13$	25	$\begin{cases}20\\2 \cdot 7 – 22 \cdot 2\|\|\end{cases}$		$5 \cdot 6$
NO_3^-	$0 \cdot 1$		$0 \cdot 02$				
NO_2^-					$\begin{cases}0 \cdot 06\\0 \cdot 04\end{cases}$	$0 \cdot 09$	$0 \cdot 066$
SO_3^{2-}					6×10^{-3}		
SO_4^{2-}	10^{-3}		$<10^{-6}$	$<10^{-3}$	$\begin{cases}6 \times 10^{-4}\\3 \times 10^{-5}\end{cases}$		10^{-5}
$S_2O_3^{2-}$					6×10^{-3}		
S^{2-}			10^{-4}		$0 \cdot 57$		$3 \cdot 5 \times 10^{-3}$
HS^-					$0 \cdot 04$		
CO_3^{2-}			6×10^{-6}		$\begin{cases}6 \times 10^{-3}\\2 \times 10^{-4}\end{cases}$		$1 \cdot 9 \times 10^{-4}$
HCO_3^-	4×10^{-3}	3×10^{-3}		$<10^{-3}$	$\begin{cases}0 \cdot 02\\0 \cdot 009\end{cases}$		
$CH_3CO_2^-$	4×10^{-3}		$1 \cdot 5 \times 10^{-4}$	$<10^{-3}$	$\begin{cases}6 \times 10^{-3}\\4 \times 10^{-4}\end{cases}$		5×10^{-3}
ClO_3^-			$0 \cdot 03$		$\begin{cases}2\\1 \cdot 14 – 1 \cdot 45\|\|\end{cases}$		$1 \cdot 1$
ClO_4^-			$3 \cdot 0$	$>10^3$	10^3		$95 \cdot 5$
OH^-	10^{-3}						
CN^-			6×10^{-4}		$\begin{cases}0 \cdot 02\\0 \cdot 01\end{cases}$		$0 \cdot 02$
PO_4^{3-}			2×10^{-4}		$\begin{cases}3 \times 10^{-4}\\10^{-4}\end{cases}$		$7 \cdot 4 \times 10^{-3}$
$H_2PO_4^-$					$\begin{cases}3 \times 10^{-4}\\5 \times 10^{-5}\end{cases}$		
HPO_4^{2-}					$\begin{cases}8 \times 10^{-5}\\3 \times 10^{-5}\end{cases}$		
$[Fe(CN)_6]_3^-$			$<10^{-6}$				$1 \cdot 6 \times 10^{-6}$

* Manufacturers' data unless otherwise quoted.
† Reference 12.
‡ Here presented as reciprocals of manufacturer's apparent selectivity constants.
|| Reference 5.
¶ Reference 13.
** Compare with Orion selectivities for the Solid State 94–17 Electrode (Table 7.)
†† Reference 88.
‡‡ Also sensitive to CNS^- and ReO_4^- (Reference 14) as well as to MnO_4^-, IO_4^- and $Cr_2O_7^{2-}$ (Reference 15).
|||| Reference 16.

Table 5 (continued)

Interfering Anion	Manufacturers' Electrodes					
	Corning Cl⁻ 476131‡	Orion Cl⁻ 92–17**	Orme's Cl⁻ Microelectrode††	Orion ClO₄⁻ 92–81‡‡	Orion ClO₄⁻ 92–81‖‖	Beckman ClO₄⁻ 39616
F^-		0·10		$2·5 \times 10^{-4}$	$2·9 \times 10^{-4}$	10^{-4}
Cl^-				$2·2 \times 10^{-4}$		10^{-4}
Br^-	2·5	$\{1·6$ $\{1·72–3·73\|$		$5·6 \times 10^{-4}$	$1·07 \times 10^{-3}$	3×10^{-3}
I^-	15	$\{17$ $\{4·4–26·7\|$		$\{0·012$ $\{0·012–0·71\|$	0·029	0·04
NO_3^-	2·5	$\{4·2$ $\{1·68–5·89\|$		$1·5 \times 10^{-3}$	$4·3 \times 10^{-3}$	$6·6 \times 10^{-3}$
SO_4^{2-}		0·14		$1·6 \times 10^{-4}$		$<10^{-6}$
S^{2-}						5×10^{-5}
HS^-			7·5			
CO_3^{2-}						2×10^{-6}
HCO_3^-	0·21	0·19	0·15	$3·5 \times 10^{-4}$	$8·8 \times 10^{-4}$	
$CH_3CO_2^-$		0·32	0·26	$5·1 \times 10^{-4}$	$1·65 \times 10^{-3}$	5×10^{-5}
ClO_3^-						0·01
ClO_4^-	5	32				
OH^-	0·4	1·0	0·72	1·0		
CN^-						2×10^{-4}
PO_4^{3-}						10^{-4}
$H_2PO_4^-$			0·09			
HPO_4^{2-}			0·97			
$[Fe(CN)_6]^{3-}$						$<10^{-6}$

TABLE 6

Percentage Error Interferences
for an Orion 92–07 Electrode

Molar Activities of Anions Involved		K_{NO_3N}*	Percentage Error†
Primary (NO_3^-)	Interferant (N^{n-})		
10^{-2}	$0·8MF^-$	6×10^{-5}	0·48
10^{-3}	$0·8MF^-$	6×10^{-5}	4·8
10^{-4}	$0·8MF^-$	6×10^{-5}	48
10^{-3}	$10^{-2}MSO_4^{2-}$	3×10^{-5}	0·3
10^{-4}	$10^{-2}MSO_4^{2-}$	3×10^{-5}	3
10^{-5}	$10^{-2}MSO_4^{2-}$	3×10^{-5}	30

* Values from Table 5.   † Using eqn(21).

23

$$\text{Percentage error} = \frac{100 \times K_{\text{NO}_3\text{N}} \times (a_\text{N})^{1/n}}{a_{\text{NO}_3}} \tag{21}$$

The effect of sulphate and fluoride on the performance of a nitrate electrode is shown in Table 6. As the interferant-nitrate ratio increases, so does the extent of interference.

Eqn(21) can thus be used to establish the error involved when measuring a particular level of primary ion in the presence of any interferant. Thus, measurements in the above nitrate-fluoride solutions comprising more than 10^{-3} M NO_3^- can be made with some confidence since fluoride offers no severe interference. This 4·8 per cent error means that the apparent nitrate activity will read high (by 4·8 per cent) for nitrate-fluoride mixtures in the proportion of 1:800. There is virtually no error (0·48 per cent) when dealing with the same ions in the 1:80 ratio.

Selectivities of Solid State Electrodes

Selectivities for solid state electrodes have been expressed in various ways. Rather than convert the published values to a common basis, the selectivities of some solid state electrodes are given in Tables 7 and 8 in their original numerical forms. This again illustrates the present chaos regarding selectivity quotations.

The methods employed for evaluating selectivities of solid state electrodes are those already described, except that Method IIC has so far been restricted[5] to a few anion liquid ion exchange electrodes. Light and Swartz[23] have, however, used a non-graphical procedure for ten anion (N) interferences (Table 8) using the Orion 94-16 sulphide electrode based on eqn(19):

$$K_{\text{SN}} \cdot a_\text{N} = \left(\exp^{\frac{E - E_1}{RT/2F}} \right) a_\text{S} - a_\text{S}' \tag{22}$$

where the suffix M is represented by S for sulphide.

24

TABLE 7

Selectivity Constants, or Ratios, for some Cation Solid State Electrodes

Electrode Type	Interfering Cation (N)																		Selectivity Expression	Reference
	Zn^{2+}	Ni^{2+}	Co^{2+}	Mg^{2+}	Hg^{2+}	Cu^{2+}	Cd^{2+}	Ca^{2+}	Mn^{2+}	Pb^{2+}	Fe^{2+}	Al^{3+}	Fe^{3+}	Na^+	K^+	Tl^+	H^+	Ag^+		
Orion (Cd^{2+})	$4 \cdot 1 \times 10^{-4}$	$0 \cdot 03$		$0 \cdot 02$	$1 \cdot 63 \times 10^{-4}$	*		$2 \cdot 24 \times 10^{-4}$	$2 \cdot 68$	$6 \cdot 08$	196	$0 \cdot 13$	$>10^5$	$3 \cdot 2 \times 10^{-8}$	$6 \cdot 7 \times 10^{-8}$	122	$2 \cdot 41$	*	K_{CdN}	24
Orion (Pb^{2+})					*	*	*						*					*	—	†
Orion (Cu^{2+})					*		*						*					*	—	†
Beckman (Cu^{2+})					*		*													
Coleman (Cu^{2+})					*		*											*	—	†
Orion (Ag^+/S^{2-})‡					*($0 \cdot 08$)														$K_{Ag\,Hg}$	23
Coleman (Ag^+/S^{2-})‡					*($1 \cdot 0$)														$K_{Hg\,Ag}$	†

* Merely listed as either "interferes" or "must be absent".
† Manufacturers' instruction manuals or data sheets.
‡ The silver sulphide electrode is a dual purpose electrode and will respond to either silver or sulphide ions.

TABLE 8

Selectivity Constants, or Ratios, for some Anion Solid State Electrodes

Electrode Type	F^-	Cl^-	Br^-	I^-	OH^-	CN^-	SCN^-	NO_2^-	NO_3^-	Interfering CO_3^{2-}
Iodide										
Beckman¶		10^6	5×10^{-3}			0.4				
Lucite‡‖	2.5×10^{-6}	10^{-5}	10^{-4}					5×10^{-6}	5×10^{-6}	
Orion¶		10^6	5×10^3			0.4				
Philips††		6.6×10^{-6}	6.5×10^{-5}			0.34				1.2×10^{-4}
Pungor‡‡		1.7×10^5	210							
Bromide										
Beckman¶		400		2×10^{-4}	3×10^4	8×10^{-5}				
Coleman¶		400		2×10^{-4}	3×10^{-4}	*				
Orion¶		400		2×10^{-4}	3×10^4	8×10^{-5}				
Philips††		6×10^{-3}		20	3×10^{-3}	25				2.3×10^{-3}
Pungor		$203^{‖‖‖}$ $200^{‡‡}$ $188\text{–}193$		7.7×10^{-3}						
Chloride										
Beckman¶			3×10^{-3}	5×10^{-7}	80	2×10^{-7}				
Coleman¶			4.9×10^{-3}	5×10^{-6}	100	*				
Orion¶			3×10^{-3}	5×10^{-7}	80	2×10^{-7}				
Philips††			1.2	86.5	0.024	400				3×10^{-3}
Pungor‖‖‖			4.95×10^{-3} 6.6×10^{-2}	2.8×10^{-3}						
Cyanide										
Philips††¶¶		2×10^{-4}	2×10^{-5}	3						3.6×10^{-4}
Orion¶		10^6	5×10^3	0.1						
Pungor¶¶				*						
Sulphide										
Orion	$<10^{-3}$	$<10^{-3}$	0 $<10^{-3}$	0 $<10^{-3}$	0 $<10^{-3}$	$<10^{-3}$	0			$<10^{-3}$

* Merely listed as either "interferes" or "must be absent".
† Manufacturers' instruction handbooks or data sheets.
‡ Quoted as KMN using equation 13.
‖ Silver iodide dispersed in Lucite 45 at 50–70 wt. per cent. Not commercially available but patent pending.
¶ Selectivity constant K_{NM}. Expressed as maximum allowable ratio of interferant to primary ion. Coleman Inc. also call this ratio the selectivity constant.
** Value shown is the ratio for a one per cent error.
†† Quoted as K_{MN} using equation 10 (modified for anions).
‡‡ Apparent selectivity ratio here defined as the ratio of $[N^{n-}]$ to $[I^-]$ necessary to yield the same cell e.m.f. by separate solution procedure.
‖‖‖ Selectivity constant K_{NM} given as the quotient of K_{AgN} and K_{AgM}.
¶¶ Philips or Pungor iodide electrodes can be used as cyanide electrodes.
*** Reference 22 using Methods IIA or IIB.
††† Reference 23 using equation 22.

HCO₃⁻	SO₄²⁻	SO₃²⁻	S₂O₃²⁻	S²⁻	PO₄³⁻	CrO₄²⁻	ClO₄⁻	C₂O₄²⁻	[Fe(CN)₆]⁴⁻	Ref.

Anion (N)

HCO_3^-	SO_4^{2-}	SO_3^{2-}	$S_2O_3^{2-}$	S^{2-}	PO_4^{3-}	CrO_4^{2-}	ClO_4^-	$C_2O_4^{2-}$	$[Fe(CN)_6]^{4-}$	*Ref.*
			10^5	*						†
	10^{-5}	3.5×10^{-4}			3.5×10^{-4}					17
			10^5**	*						†
		7.1×10^{-4}				3.7×10^{-3}				†
	3.2×10^7				4.8×10^5		1.6×10^6		3.3×10^3	18
				*						†
				*						†
				*						†
		1.5		*		1.6×10^{-3}				†
	10^8‖‖									19
										20
										21
		0.01		*						†
				*						†
		0.01**		*						†
		60		*		1.8×10^{-3}				†
	4.95×10^5									19
										20
		2.2×10^{-3}		*		1.4×10^{-2}				†
				*						†
				*						†
$<10^{-3}$	0 / $<10^{-3}$	$<10^{-3}$	$<10^{-3}$			0	0			*** / †††

The potential, E, is first measured in Na_2S (10^{-2} M) and then the potential E_1 taken for [Na_2S (10^{-2} M) + NaN (10^{-2} M)] solution, both solutions being molar with respect to sodium hydroxide.[23] The high level of sodium hydroxide liberates most of the sulphide complexed in HS^- or H_2S and also fixes the total ionic strength.

Hseu and Rechnitz[22] have studied eight anion interference effects on the same Orion sulphide electrode in mixed solutions (Methods IIA or IIB). Their e.m.f./activity graphs coincided with the calibration graph due to sulphide alone. No selectivities could thus be measured; in fact they are infinitely small.[22] This contrasts with the finite selectivities found[23] for chloride, bromide, iodide, and sulphate ions using eqn(22) (Table 8).

All ion selective electrodes are similar in operative principle in that chemical exchange equilibria occur at the analytical sample-membrane interface. They differ merely in the mechanism involved when the primary ion, that is, the one of actual analytical interest, and the interfering ion(s) gain access to the all-important membrane. Any meaningful selectivity of liquid ion exchange membrane (and glass) electrodes depends on the following generalized equilibrium scheme lying well to the right:

$$RN + M^{n+} \rightleftharpoons RM + N^{n+} \tag{23}$$

For liquid ion exchange electrodes, R represents one of many large mobile organic exchange "sites" in the membrane. The interference is rather different for solid state electrodes. Interferences are caused in several ways; for example, problems arise when the membrane material reacts with interferant to give a soluble complex. Citrate (Ct) is a classical case for the lanthanum fluoride electrode:

$$LaF_3(s) + Ct^{3-}(aq) \rightleftharpoons LaCt(aq) + 3F^-(aq) \tag{24}$$

This additional aqueous fluoride, arising in the membrane, causes the electrode to read "high" when the actual analytical solution itself is dilute with respect to fluoride.

28

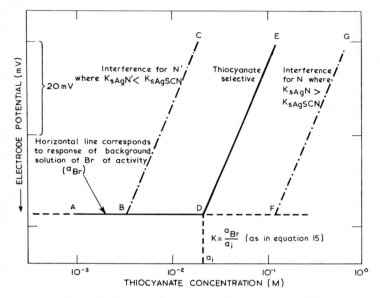

Fig. 11. *Effect of thiocyanate interferant on a bromide electrode (modified from Reference 7).*

Another common form of interference arises when the interfering ion actually reacts at the membrane surface to form a new, and insoluble (not soluble as with the fluoride electrode example) compound.[7] Thus, thiocyanate reacts with the silver bromide component of an Orion 94-35 bromide electrode:

$$SCN^- + AgBr(s) \rightleftharpoons AgSCN(s) + Br^- \qquad (25)$$

This electrode can tolerate thiocyanate, but only to a certain level, when there is an abrupt change in potential response as seen using Method IIB (Fig. 11). Indeed, the electrode has really converted to a thiocyanate electrode with the expected slope of about 59 mV per decade. This interfering film of silver thiocyanate does not cause permanent damage, and is readily removed by wiping, or using a toothbrush and toothpaste.[7] Fresh silver bromide becomes exposed, and the electrode reverts to its former rôle as a bromide-sensitive electrode. The abrupt onset of interference for Orion solid state electrodes (Fig. 11) has been contrasted[7] with the typically gradual interference patterns for liquid membrane electrodes (Figs. 7 and 9). However, this feature is not apparent in a number of interference plots for either Pungor or Lucite heterogeneous solid state electrodes.

If thiocyanate did not interfere, the e.m.f. response would be constant so as to give a line ADF and not ADE as found. The selectivity is then taken as the reciprocal of the value found using eqn(15). Other interferences are included in Fig. 11 to show (ideally) the expected earlier (ABC) and delayed (AFG) onsets of interferences for anions whose silver salts are respectively less soluble, and more soluble, than silver thiocyanate.

Reaction 25 can only proceed to the right, and thereby cause interference, if the ratio of a_{SCN} to a_{Br} exceeds a particular value. This value is simply given by the quotient of their respective solubility products, $K_{s_{AgSCN}}/K_{s_{AgBr}}$, namely to $10^{-12}/3 \cdot 3 \times 10^{-13} = 3$. Beyond this value the interfering silver thiocyanate film begins to cover the surface of the silver bromide.

This concept leads to some important consequences. Firstly, selectivity values for solid state electrodes should actually be numerically constant, unlike those for liquid membrane electrodes. The figures presently available support this idea (Tables 7 and 8). However, it would be most interesting to take say a bromide electrode and follow methods IA, IB, IIA, IIB and IIC for several different anion interferants, so as to compare experimental and calculated selectivities. Secondly, the selectivities can be found from published solubility product data, and without the need for experimentation (Table 9). This is just not feasible for liquid membrane electrodes.

The correlation varies between good (chloride electrode) and poor (cyanide electrode). Interferences calculated from

TABLE 9
Experimental and Calculated Selectivities for Solid State Electrodes

Interfering Anion	Maximum allowable ratio of N/M quoted for Orion Electrode				Solubility Product, Ks, of Relevant Silver Salt (Reference 25)
	94–06 CN⁻	94–17 Cl⁻	94–35 Br⁻	94–53 I⁻	
OH^-	—	80(110)	3×10^4 (6×10^4)	—	AgOH 2×10^{-8}
I^-	$10^{-1}(1.2 \times 10^{-2})$	5×10^{-7} (5×10^{-7})	2×10^{-4} (2.6×10^{-2})	—	AgI 8.5×10^{-17}
Br^-	$5 \times 10^3(47)$	3×10^{-3} (2×10^{-3})	—	5×10^3 (3.8×10^3)	AgBr 3.3×10^{-13}
Cl^-	$10^6(2.4 \times 10^4)$	—	400(510)	10^6 (2×10^6)	AgCl 1.7×10^{-10}
CN^-	—	2×10^{-7} (4×10^{-5})	8×10^{-5} (2.1×10^{-2})	0.4 (82)	AgCN 7×10^{-15}

Parenthesised values calculated from $K_{s_{AgN}}/K_{s_{AgM}}$

solubility products can therefore be misleading, for various reasons. Thus, a low level of silver ions should interfere[7] with measurements of copper, cadmium and lead ions owing to the low solubility of silver sulphide ($K_{s_{Ag_2S}} = 10^{-51}$). However, the reaction necessary to form the interfering Ag_2S film is very slow. As a result of this favourable kinetic effect, up to 10^{-7} M total silver can be tolerated in mixed solutions containing about 10^{-3} M copper, cadmium or lead.[7]

Thirdly, it is possible to predict the order of selectivity, for example, $I^- > Br^- > Cl^-$, for silver halide electrodes. This is because equilibria, such as eqn(25), are shifted furthest to the right by the interferant whose silver salt is the least soluble. Thus, the chloride electrodes (except the Philips model) suffer quite massive interferences from iodide, whereas the four commercial iodide electrodes can withstand equally massive amounts of chloride (Tables 8 and 9). This means that silver halide electrodes are quite ineffective when anions are present whose silver salts are less soluble than the sensor silver salt comprising the membrane of the electrode.

31

The discrimination for chloride by the four iodide electrodes is essentially the same. This is interesting since the Pungor and Lucite electrodes are each fabricated quite differently from the Orion model and presumably the Philips electrode, although all four contain silver iodide. The Lucite and Philips selectivities are really expressed in the reciprocal sense to the Orion and Pungor electrodes. However, when "converted" to the Orion-Pungor definition, the chloride selectivities for the four iodide electrodes all lie between 10^5 and 10^6. This means that these iodide electrodes can tolerate very large excesses of chloride indeed before going out of commission. Except for the Pungor iodide electrode, there is also reasonable agreement concerning the level of bromide interference, and which as expected, is relatively more severe than chloride (Table 8).

These maximum allowable selectivity ratios (like the selectivity constants discussed under liquid membrane electrodes) are very useful parameters for the analyst. For example, it is possible to calculate the maximum permitted pH at which a 10^{-4} M potassium chloride solution may be measured with a Coleman electrode. Thus, the maximum hydroxide ion level is $10^2 \times 10^{-4} = 10^{-2}$ M, that is, pH 12. Large selectivity ratios are desirable, small values undesirable, although unsuspected kinetic effects may offset unfavourable values in practice.

Other Types of Interferences

Foreign ions affect the ionic strengths of solutions to varying degrees, which in turn affect the activity of the primary ion. Many other interferences are due to complexation of the primary ions. Irrespective of whether the complexes are soluble or precipitates, the nett result is a lower activity of primary ion and an incorrect low, or high, e.m.f. response.

3

GLASS ELECTRODES

Glass electrodes[26,27] selective to hydrogen ions have been standard items of laboratory equipment since the late 1930s, although their response deviates from the Nernst equation in alkaline solutions.[28] This deviation, known as the "alkaline error", increases with increasing aluminium oxide content of the glass and is due to the electrode responding to alkali metal ions.

More recently, glass membrane electrodes with modified glass composition have become available. These frequently respond selectively to cations other than hydrogen ions, so that the glass electrode represents a family of glass compositions in which the composition of the classical pH electrode is merely an extreme member.

Origin of the Glass Electrode Potential

It is now generally accepted that the operation of a glass electrode depends on an ion-exchange process.[29] A cross-section through the membrane of a functioning glass electrode comprises several discrete regions and interfaces:

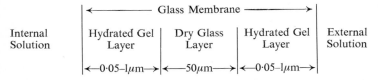

The dry glass layer constituting the bulk of the membrane's thickness is sandwiched between two much thinner hydrated layers which are essential for the proper functioning of glass electrodes. This is probably because the diffusion of ions in the glass is considerably boosted by hydration, the diffusion coefficients of univalent cations being about 10^3 times greater in hydrated glass than in dry glass.

The glass membrane potential, E_M, represents contributions from both diffusion and phase-boundary processes:

Internal Solution	Hydrated Gel Layer	Dry Glass Layer	Hydrated Gel Layer	External Solution
$\longleftarrow E_A \longrightarrow$		$\longleftarrow E_D \longrightarrow$		$\longleftarrow E_B \longrightarrow$

where E_A is the phase-boundary potential at the internal solution/membrane interface, E_D the diffusion potential in the glass membrane and E_B the phase-boundary potential at the external solution/membrane interface, that is,

$$E_M = E_A + E_D + E_B.$$

To all intents and purposes, the diffusion potential, E_D, is constant for a given electrode. As the internal solution is kept constant in glass electrodes, the phase-boundary potential, E_A, at the internal solution/membrane interface is also constant. Hence the potential given by the glass electrode depends only on the phase-boundary potential, E_B, at the external analytical solution/membrane interface and which in turn is related to ionic activity by the Nernst equation.

Although hydrogen ions undergo exchange across the solution/hydrated layer interface, these ions do not penetrate the glass membrane. This was elegantly demonstrated by coulometric experiments involving prolonged electrolyses in glass electrode bulbs filled with tritium-labelled sample solutions.[30] If hydrogen ions penetrate the membrane, some appreciable fraction of the total quantity of electricity passed during electrolysis should be accounted for by the transport of hydrogen ions, and thus tritium should be found on the other side of the membrane. In fact, the quantity of tritium found on the non-labelled side of the membrane never exceeded the natural tritium content of the outside

34

solution, even after 20 h of electrolysis at elevated temperatures. This experiment conclusively disproved the long standing view that the glass membrane was completely permeable to hydrogen ions.

Obviously, some small but finite current must flow during the potentiometric measurement, and charge must be transported across the entire glass membrane system. This charge can be transferred across the solution/hydrated layer interface by ion-exchange, and within the hydrated layer by diffusion.

How, then, is charge carried across the dry glass portion of the membrane? All available experimental evidence indicates an ionic mechanism, involving the cationic species of lowest charge available in any given glass. No single sodium ion (for sodium silicate glass) moves through the entire thickness of the dry glass layer, but the charge is regarded as being transported by an interstitial mechanism, where each charge carrier merely needs to move a few atomic diameters before transferring its energy to another carrier.

Potentiometric Behaviour or Glass Electrodes

Subject to the limitation that potential measurements of cells incorporating the high resistance glass electrode require measuring devices with very high input impedance, the potentiometric behaviour of a pH glass electrode in a solution containing hydrogen ions as the only cations may be described by the Nernst equation:

$$E = E^\circ + \frac{RT}{F} \ln a_{H^+}$$

or

$$E = E^\circ - \frac{2\cdot303\,RT}{F}\,\mathrm{pH} \tag{26}$$

where a_{H^+} is hydrogen ion activity; E the electrode potential; and E° the standard electrode potential.

Eqn(26) is a linear relation between E and pH of slope $-2\cdot303RT/F$ and E becomes more negative by 59 mV for each unit increase in pH at 25°C.

However, as mentioned, the response deviates from ideal linearity when alkali metal ions are present in the solution at pH > 10. Thus, ion-exchange occurs[30] between the alkali metal ions (N^+) in solution, and the hydrogen ions (H^+) in the glass membrane:

$$H^+{}_{glass} + N^+{}_{solution} \rightleftharpoons N^+{}_{glass} + H^+{}_{solution} \qquad (27)$$

The selectivity coefficient, K_{HN} is simply given by:

$$K_{HN} = \frac{a_H\, a'_N}{a'_H\, a_N} \qquad (28)$$

where a and a' represents solution and glass activities, respectively.

From this it can be shown[31] that the equation describing a glass electrode response in a solution containing both hydrogen ions and any one alkali metal ion is:

$$E = E^\circ + \frac{2\cdot303RT}{zF} \log\,(a_H + K_{HN}a_N) \qquad (29)$$

If a_H is much greater than $K_{HN}a_N$, the electrode exhibits a hydrogen ion function, while if $K_{HN}a_N$ is greater than a_H, the electrode exhibits a metal ion response.

Thus, if a pH electrode is required, the glass composition is designed to have the widest possible range of pH response, that is, K_{HN} is very small. Indeed for all pH glasses studied, K_{HN} lies between 10^{-1} and 10^{-15}.

If a specific metal ion electrode is required, the glass composition is chosen to show a metal ion function at the lowest possible pH, i.e. K_{HN} is large.

A more general form of eqn(29) for a binary mixture of univalent cations is:

$$E = E^\circ + \frac{RT}{F} \ln\,(a_M^{1/n} + K_{MN}a_N^{1/n})^n \qquad (30)$$

where a_M and a_N represent the molal activities of the cations M^+ and N^+, and K_{MN} and n are constants depending upon the glass composition.[32,33]

Effect of Glass Structure on Selectivity

Developing glasses for glass electrodes still tends to be more art than science. Nevertheless, painstaking study has shown that some formulations are better than others. With regard to hydrogen-ion sensitive electrodes the most common glasses employed are alkali metal silicates. A pure silica glass has the structure:

$$-\overset{|}{\underset{|}{Si}}(\text{IV})-O-\overset{|}{\underset{|}{Si}}(\text{IV})-$$

Because no charged sites are available for ion-exchange this structure shows no electrode properties. However, when an alkali metal oxide is incorporated into the glass, the lattice is disrupted, yielding some silicon-oxygen sites of the form:

$$-\overset{|}{\underset{|}{Si}}(\text{IV})-O^-$$

This type of site exhibits a high specificity for hydrogen ions over other cations.

To realize cation selectivity relative to hydrogen ions it is essential to introduce a structural element into the glass lattice in a co-ordination state higher than its oxidation state. For example, when an element R, in oxidation state $+3$, is introduced into glass in four-fold co-ordination in place of silicon (IV), the structure becomes:

$$[-\overset{|}{\underset{|}{R}}(\text{III})-O-\overset{|}{\underset{|}{Si}}(\text{IV})-]^-$$

in which the replacement has resulted in a site capable of cation exchange. Thus, by incorporating alumina into alkali silicate glasses, cation-responsive glasses are obtained.

37

Conversely, it has been argued that substitution of phosphorus(v) for aluminium(iii) might result in anion responsive glass.[32,33]

If an element in oxidation state $+4$ and four-fold co-ordination, for example, zirconium(IV) is introduced into the glass, so as to realize the structure

$$-\overset{|}{\underset{|}{Zr}}(\text{IV})-O-\overset{|}{\underset{|}{Si}}(\text{IV})-$$

then, as with pure silica glass, there is no net charge and no ion-exchange property. However, because zirconium(IV) is capable of existing in six-fold co-ordination the possibility of the structure

$$[-\overset{|}{\underset{/}{\overset{/}{Zr}}}(\text{IV})-O-\overset{|}{\underset{|}{Si}}(\text{IV})-]^{2-}$$

arises, resulting in a local site bearing a double negative charge. This can function as an ion exchange site for two singly charged cations, or one doubly charged cation, and indeed this zirconium glass has been found to exhibit electrode properties.

Having seen how selectivity for monovalent cations over hydrogen ions arises, it is pertinent to consider the selectivity shown by some glasses for one particular metal ion over another. The majority of cation-selective glasses are alkali metal aluminosilicates, the most common being sodium aluminium silicate. Within this glass system a wide variety of particular selectivities for the various cations is found, ranging from electrodes highly specific for Na^+ to those usefully selective for K^+, as well as glasses having some application for Li^+, Rb^+, Cs^+, Ag^+, Tl^+ and NH_4^+ (and possibly even bivalent cations in certain limited applications). In fact, all the principal features of present-day cation-sensitive glass electrodes are exemplified within the Na_2O-Al_2O_3-SiO_2 system. Indeed, the modified glass need

38

not, and does not, comprise the cation to which it is specifically sensitive. Some of the useful glass compositions within this system are shown in Table 10.

TABLE 10

Examples of Compositions for Cation Sensitive
Glass Electrodes (Reference 34)

Principal Cation to be Measured	Percentage Glass Composition			Approximate Selectivity
	Na_2O	Al_2O_3	SiO_2	
Na^+	11	18	71	$K_{NaK}3.6 \times 10^{-4}$ at pH 11
K^+	27	5	68	$K_{KNa}0.05$
Ag^+	28.8	19	52.2	$K_{AgH}10^{-5}$
Ag^+	11	18	71	$K_{AgNa}10^{-3}$

Practical Aspects of Glass Electrode Use

The resistance of the glass electrode is very high, so that the input resistance of the measuring device for measuring the voltage of any cell incorporating the glass electrode must likewise be high; this is to avoid drawing an unduly high current during a measurement. For this reason, the instrument frequently used to measure the voltage of cells such as this is a vacuum tube voltmeter in which the potential of the cell controls the current flow, the meter being calibrated in volts or in pH units; in the latter case, it is called a pH meter.

Before use, the glass electrode must be well soaked. This ensures that the diffusion potential which governs the electrode response depends solely on the relative mobilities of ions in the thin swollen hydrated glass layer. The glass electrode owes its amazing versatility to the constant self-regeneration of these hydrated zones.

With regard to practical application, activity coefficients and ionic strength of the sample solution will affect the potential developed by any ion-sensitive electrode, with glass and non-glass membranes. The presence of any complexing species will influence the activity of desired species, even though their total concentration remains constant. Without proper allowance for this electrode property, analytical methods giving total analytical concentrations, for example flame photometry, will be at considerable variance with those obtained by ion-sensitive electrodes.

If the activity coefficients for an ion are available, and if the ionic strength of a medium is known, it is a relatively simple matter to relate activities to concentrations and vice versa (p. 5). The situation becomes more complicated, however, when interfering ions are present, especially if the interfering ion has a different charge from the ion of interest. The effect of ionic strength on the activity coefficients of the two ions will usually be different, and the task of converting the measured cell e.m.f. to ion concentrations becomes difficult.

Care must also be exercised in selecting a suitable reference electrode. The most common type used for pH measurement is a calomel electrode with a saturated potassium chloride salt bridge. The arrangement with the calomel reference electrode combined within the same module as the glass electrode (Fig. 12) is especially neat. However, the calomel electrode will obviously be unsatisfactory in combination with a potassium sensitive electrode, because leakage of potassium chloride from the salt bridge will increase the activity of potassium in the sample solution. In some cases rubidium chloride has been used as the salt bridge electrolyte.

To date, only cation sensitive glass electrodes are available, although OH^- is indirectly assayed with the classical pH glass electrode. Furthermore, neither modified glass nor the newer non-glass electrodes (except for the silver sulphide electrode) are as sensitive as the classical pH type which detects H_3O^+, or OH^- at about the 10^{-14} molar level.

Despite these limitations, cation-sensitive electrodes have been extensively used in diverse applications, and the fact that such glass electrodes do not respond to complexed cations makes them very suitable for complex formation[35,36] and ion association[37] studies. They have been used for reaction

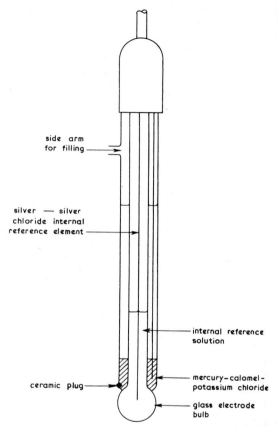

side arm
for filling

silver — silver
chloride internal
reference element

internal reference
solution

ceramic plug

mercury-calomel-
potassium chloride

glass electrode
bulb

Fig. 12. A combined calomel reference-glass electrode.

rate measurements[38] and as electrodes in potentiometric titrations of K^+, Cs^+, Rb^+, NH_4^+ and Ag^+ ions with tetraphenylborate solution.[39]

A sodium-sensitive glass electrode has been reported as being useful for the determination of sodium in water in the range 0·005 to 25 p.p.m.[40] Also, the electrode has been used for determining sodium (1-50 μg litre^{-1}) in high purity waters (e.g. condensed steam) from power stations.[41]

41

4

Because of their non-destructive operation, their rapid response, and the ease of fabrication and miniaturization, cation-sensitive electrodes are potentially useful in clinical applications, and indeed have been used for measuring Na^+ and K^+ ion activities in flowing blood streams and static body fluids.[42,43] The sodium ion selective electrodes are useful in the food industry, examples being the determination of salt in bacon[44] and sodium in egg yolk.[45] Cation selective electrodes are also claimed to function satisfactorily in systems containing non-aqueous solvents.[46,47,48] Doubtless, applications will increase, particularly for on-stream analysis of flowing solutions, although some means of standardizing the electrode assembly at intervals should be provided. Additionally, improvements and developments can be expected in electrode response. For example, the reported[49] sensitization incurred by using a film of immobilized urease on the glass bulb is likely to be exploited. Such sensitization consists of a linear response to lower cation concentrations without affecting selectivity order, and freedom from interference by alkaline earth cations.

4

GENERAL EXPERIMENTAL PROCEDURES WITH SELECTIVE ION SENSITIVE ELECTRODES

The new selective ion sensitive electrodes, irrespective of type, are used in the same way as the classical glass electrode. The basic experimental set-up incorporates two other units— a reference electrode to complete the electrochemical cell, and a voltmeter with high input impedance to measure the e.m.f. response (Fig. 13). A pH meter of appropriate sensitivity is usually suitable for this purpose. A chart recorder attachment can be a very useful addition. Some manufacturers have marketed new and elegant digital pH meters and voltmeters as well as such things as double junction reference electrodes and activity standard solutions for electrode calibration. No extensive coverage is made of this ancillary equipment.

Fundamental Requirements of Selective Ion Electrodes

Shatkay[50] has compiled a most important list of the essential requirements for selective ion sensitive electrodes. Thus, any electrode should:

- (i) either be commercially available, or at least be capable of "home-construction";
- (ii) be easily handled;
- (iii) be robust, durable and not poisoned;
- (iv) give reproducible and meaningful results within the everyday range of ion activities (or concentrations). These are commonly $10^{-2} \rightarrow 10^{-4}/10^{-5}$ M;

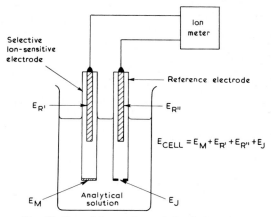

Fig. 13. *Essential experimental circuit for using any selective ion-sensitive electrode.*

(v) be specific, that is, measure the correct activities of the primary ion irrespective of any other ions present;

(vi) have low response times so that a constant e.m.f. is attained within minutes at most.

Finally, it is helpful, but not essential, if the e.m.f./activity character is Nernstian.[50]

Many of the new electrodes match up to these rather stiff demands.

General Characteristics of Selective Ion Sensitive Electrodes and Activity Versus Concentration

There are certain outstanding features of selective ion sensitive electrodes.

(i) They usually exhibit short response times to activity (or concentration) changes; this is of special importance for continuous process control by on-stream monitoring.

(ii) Little or no pretreatment is necessary before use, and the sample solution can be opaque, turbid, coloured and contain protein.

44

(iii) A meaningful e.m.f. response may be given over a concentration (or activity) range far wider than is possible by any other single physical device. Thus, the fluoride electrode spans about six activity decades (*ca.* $10^0 \rightarrow 10^{-6}$ M F$^-$) while the sulphide electrode probably runs to about twenty decades. However, quotations for sulphide detection around 10^{-24} M levels are patently absurd. These impressive operational capabilities have their drawbacks. The precision of single ion measurements is restricted to around ± 2 per cent for singly charged ions and about ± 4 per cent for doubly charged species. The electrodes, like the pH glass electrode, may be employed in a direct manner following calibration, or provided the activities are fairly high ($> 10^{-3}$ M) for classical potentiometric titrations.

(iv) Most important, electrodes sense ionic activities (p. 5) not ionic concentrations, although in dilute solutions (Figs. 4 and 14) the terms are synonymous. Because of this, a knowledge of activity parameters is often more desirable, or essential, than molar or gram concentrations, although both concepts have their merits.

Any divergence between the activity and concentration is related to the ionic strength and possibly to complexation. These electrodes respond only to free, that is, to non-complexed ionic species. It is proper, however, to mention that in aqueous solutions all ions exist as aquo species. For example, sulphide, as S^{2-}, undergoes various degrees of complexation as the pH falls, to give HS$^-$ and H$_2$S; neither is "sensed" by the sulphide electrode. The total sulphur concentration, however, remains the same, irrespective of pH, and the electrode will unequivocally indicate a low sulphide activity or concentration level. This apparent drawback can be turned to profitable use in that many hitherto difficult complexation studies have become possible.

The activity of any ion depends on the total ionic strength (μ) of a solution as already discussed (p. 5). Ionic strengths are easily calculated from the molar concentrations of individual components in a solution, and the activity of one species will drop if the concentration of other species is increased. Numerous equations are available for calculating activity coefficients, which in turn permit very simple calculations of single ion activities from $a = fc$. These μ–f working equations are arithmetically tedious. However, a "once and for all" computer simulation trace is easily obtained and for six such equations,[5,6,51,52] is shown in Figs. 4 and 14. A little care is needed with the equation involving the effective ionic radius, R, which depends on the actual ion in question. These working plots[53] are essentially identical for $\mu < 10^{-3}$, but two distinct patterns appear at higher ionic strengths. The choice of equation is personal, but the one associated with curve 2 (Figs. 4 and 14) has been extensively used.[53] The activity of divalent ions is more adversely affected by changes in μ than are monovalent ions.

In order to evaluate selectivity constants in mixed solutions, for example, Method IIA (p. 14), it is necessary to know the activities of primary and interferant ions at the extrapolated break (Figs. 6 and 7).

The procedure for estimating calcium and magnesium activities in a mixture is illustrated in Table 11. The calcium ion activity at the break is 6.5×10^{-5} M (Fig. 7), and the corresponding magnesium ion activity is 2.67×10^{-3} M (Table 11). Thus, K_{CaMg} is just the ratio $6.5 \times 10^{-5}/2.67 \times 10^{-3}$; that is, 0.024 (Table 2). A similar set of activities is calculated for calcium-sodium mixtures in Table 12. When these calcium activities (Table 12) are plotted on semilog paper against the corresponding e.m.f. values, the required calcium activity at the break is (by sheer chance) also the same value as before, namely 6.5×10^{-5} M, wherever the sodium activity is 0.55 M. Therefore,

$$K_{CaNa} = 6.5 \times 10^{-5}/(0.55)^2 = 2.1 \times 10^{-4}$$

(Table 2).

46

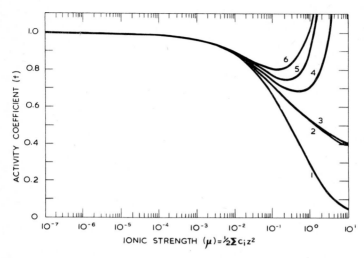

Fig. 14. Computer simulations of various activity coefficient-ionic strength equations for univalent ions. (Each plot represents 800 μ–f pair values.)

TABLE 11

Calculations of Activities in a M̲i̲ˣ̲

Page 47. Legend to Fig. 14.

Curve 1: $\log f = -Az^2\sqrt{\mu}$

Curve 2: $\log f = -Az^2\sqrt{\mu}/(1 + \sqrt{\mu})$

Curve 3: $\log f = -Az^2\sqrt{\mu}/(1 + 0.329\ R^*\sqrt{\mu})$

Curve 4: $\log f = -Az^2\left(\dfrac{\sqrt{\mu}}{1 + 1.5\sqrt{\mu}} - 0.2\mu\right)$

Curve 5: $\log f = -z^2\left(\dfrac{A\sqrt{\mu}}{1 + \sqrt{\mu}} - 0.2\mu\right)$

Curve 6: $\log f = -z^2\left(\dfrac{A\sqrt{\mu}}{1 + 1.5\sqrt{\mu}} - 0.2\mu\right)$

* For plot 3, R taken as 3 Å.

TABLE 12

Calculations of Activities in a Mixture of Monovalent and Divalent Ions[53]

(Here 10 ml of 2MNaCl is added to 10 ml of six different $CaCl_2$ solutions ranging from $10^{-1} \rightarrow 10^{-6}$ M.)

C_{Ca}^*	C_{Na}^*	Ionic Strength† (μ)	Activity Coefficients		Activities $(c \times f)$‖	
			$f_{Ca^+}^+$	f_{Na}¶	a_{Ca}	a_{Na}
5×10^{-2}		$1 \cdot 15$			$4 \cdot 5 \times 10^{-3}$	
5×10^{-3}		$1 \cdot 015$			$4 \cdot 5 \times 10^{-4}$	
5×10^{-4}		$1 \cdot 001$			$4 \cdot 5 \times 10^{-5}$	
5×10^{-5}	$1 \cdot 0$	$1 \cdot 000$	$0 \cdot 09$	$0 \cdot 55$	$4 \cdot 5 \times 10^{-6}$	$0 \cdot 55$
5×10^{-6}		$1 \cdot 000$			$4 \cdot 5 \times 10^{-7}$	
5×10^{-7}		$1 \cdot 000$			$4 \cdot 5 \times 10^{-8}$	

The *, †, ‡ and ‖ are as for Table 11.
¶ Obtained from μ-f plot of Fig. 14 (Curve 2).

Measurement Procedures with Selective Ion Electrodes

Frequently, as with the rates and extents of chemical and biochemical reactions, activities are of relevance. Conversely, concentration is often of greater significance, as with cyanide effluents. In either case the electrode must be calibrated. A good quality pH meter, or a specific ion meter, is suitable for this purpose.

Specific ion meters analogous to pH meters are sophisticated devices representing an advance parallel with the new electrodes themselves. The Orion 400 series of ion meters are light, portable, accommodate electrodes of up to 10^4 megohms resistance, incorporate a slope corrector for simple calibration and can be run for a year at about four hours per day without battery replenishment. They are as simple to use as pH meters and can give direct readings of ionic activities, concentrations, pH and millivolts. One drawback is that readings are only possible to ± 1 mV, and for precise work, either the Beckman Research Model pH meter or the Orion or Corning digital pH meters reading to $\pm 0 \cdot 1$ mV are essential.

Stock solutions for dilution can easily be prepared for calibration purposes using known weights of A.R. compounds and doubly, or triply, distilled water. Alternatively, ready-made activity stock solutions for most electrodes can now be purchased. (About $7 per U.S. pint.)

Although stirring of solutions is strongly recommended (except for sulphide), the thermal output of some magnetic stirrers can sometimes cause errors (re Nernst equation where $\pm 5°C \sim \pm 1 \, mV$). However, a cork pad under the sample beaker will offset this undesirable effect. Known and unknown solutions should be stirred and be at the same temperature.

Some standardizations are made difficult owing to oxidation of samples, for example, sulphide is easily oxidized to thiosulphate and sulphate. This oxidation may be reduced but not completely prevented by using an inert gas stream. Dilutions must be performed with freshly boiled water so as to expel dissolved oxygen. In other cases, for example, with dilute copper standards, contamination arises through using water conveyed in copper pipes.

Various procedures are available for estimating ions in solutions with these electrodes. However, this is not possible until the particular electrode has been calibrated against solutions of known activity or concentration. Potentials developed by an electrode can change as much as 10–20 mV/ day depending, for example, on the reference electrode. For high precision work, several calibration checks may be essential each day. Some of the various methods developed for measuring the activity, and concentration, of ions will now be outlined.

Activity Measurements

For calibration, the particular selective ion sensitive electrode-reference electrode pair is placed in a series of solutions, each of known activity, and starting with the most dilute solution, the e.m.f. read off the coupled up millivolt meter.

49

All these e.m.f. readings are then plotted against corresponding activities on semilog paper. It would be particularly useful if such data could be universally presented exactly as in Figs. 2, 3 and 5 for example.

Once calibrated against standard activities the electrode responds to the activity of a particular ion regardless of the remaining sample ion composition or its total ionic strength, *provided*, of course, interferant species are absent; for example, iodide in the case of the Corning nitrate electrode.

The procedure is somewhat simpler when using an Orion specific ion meter, since two known activity standards are sufficient and no calibration plot is necessary. These instruments, like the pH meter, carry a log scale for direct readout of activity.

One aim of the two-year joint National (U.S.A.) Bureau of Standards and Science Apparatus Makers Association (U.S.A.) research programme will be to establish consistent pIon scales as for pH. The U.K. SIRA Institute has also undertaken an assessment programme.

Concentration Measurements of Uncomplexed Ions

The actual method used to estimate uncomplexed, that is, free ions depends on sample make-up.

Method A. Pure Solution Method

When a sample consists largely of one salt, for example, sodium fluoride, the previous activity—e.m.f. calibration data can easily be converted to a concentration—e.m.f. plot, since f is readily obtained from one of the f–μ plots and $C = a_{ion}/f_{ion}$. Such calibrations are rarely linear over a wide concentration range as illustrated (broken lines) in Figs. 2 and 3.

The same procedure is used for mixed solutions where the ratio of the primary ion varies directly with the other ion(s). Thus, to estimate potassium in sea-water, calibration

50

standards are made by diluting synthetic potassium sea-water samples with distilled water. This calibration can then be used to estimate potassium in sea-water.

Method B. Constant Ionic Background Method
When the analytical sample contains a high, but essentially constant, level of ions besides the one of interest, then calibration standards of similar composition are employed. This ensures that the standards, and unknowns, have the same total ionic strengths.

Method C. Ionic Strength Adjustor Method (I.S.A.M.)
In many cases, the samples can exhibit considerable overall variation in composition. Adding an "ionic strength adjustor" to these samples, and the calibration solutions, tends to bring all the solutions to a common total ionic strength. It is further assumed that the ultimate ionic strength value arises in the I.S.A., and not through the original ionic levels of unknown and standards. Virtually any substance may serve as an I.S.A., provided that substance does not complex with the primary ion (M), of interest, and also that $K_{M.I.S.A.}$ is negligible.

Measurements of Total Concentration

The previous methods give the concentrations (or activities) of free, that is, uncomplexed ions in analytical solutions. These values may be quite different from the total concentrations (that is, free + complexed species). Two very powerful, elegant, and yet simple-to-use methods are now available for total concentration measurements.[54-56] Electrode calibration is not necessary and only two potentials need be taken. Both methods require a change in the concentration of sample species, $M^{z\pm}$. For known addition, more $M^{z\pm}$ is introduced, whereas in known subtraction the level is lowered by adding a complexing agent. There are no complications over ionic strengths as in the previous cases.

51

Known Addition Method

This is sometimes referred to as the spiking or known increment method. The potential, E_1, is first taken for the unknown solution volume, V_0, and total molar concentration, C_0,

$$E_1 = E° \pm 2{\cdot}303 \frac{RT}{zF} \log x_1 f_1 C_0 \qquad (31)$$

where f_1 is the activity coefficient, and x_1 the fraction of free, that is, uncomplexed ions. Next, and finally, a small volume, V_A, of a standard concentration, C_S, and about 100 times more concentrated than the expected C_0 value (but about 100 times less than the volume of sample) is pipetted into that sample. The new potential, E_2 (where $E_2 > E_1$), taken after stirring, is given by:

$$E_2 = E° \pm 2{\cdot}303 \frac{RT}{zF} \log [x_2 f_2 C_0 + x_2 f_2 C_\Delta] \qquad (32)$$

where the C_Δ term is the change in concentration after the known addition stage, and therefore:

$$C_\Delta = \frac{V_A C_s}{V_0} \qquad (33)$$

Combination of eqns (31) and (32) gives

$$E_2 - E_1 = \Delta E = \pm 2{\cdot}303 \frac{RT}{zF} \log \left[\frac{x_2 f_2 (C_0 + C_\Delta)}{x_1 f_1 C_0} \right] \qquad (34)$$

An essential principle of the method is that $x_1 \sim x_2$ and $f_1 \sim f_2$.
Hence, eqn(34) is considerably simplified,

$$\Delta E = S \log \left(\frac{C_0 + C_\Delta}{C_0} \right) \qquad (35)$$

"where S represents the Nernstian factor $\pm 2{\cdot}303\ RT/zF$."
The following advantages of known addition have been listed.[56]

(i) Although the electrode still reads out activity, the known addition gives the total concentration, C_0; that is, complexed and uncomplexed ions.
(ii) The electrode requires no calibration.
(iii) No calibration curves have to be plotted.
(iv) The method is fast and simple in operation.
(v) Only one standard stock reagent is needed; it is long lasting, since V_A is small.
(vi) It is the only available method, when using an ion selective electrode, for determining the total concentration of an ion if a large excess of complexing agent is present.

Calculation of C_0 from Known Addition Data
Three general procedures are available to estimate C_0 from the measured e.m.f. values.
(i) The experimental ΔE and C_Δ values can be directly substituted into eqn(35).
(ii) The Orion 407 Specific Ion Meter carries special, separate, known addition and subtraction scales for the direct read-out of total concentration.
(iii) Equation (35) can be easily rearranged to:

$$\frac{C_0}{C_\Delta} = \left[\frac{1}{(\text{antilog } \Delta E/S) - 1} \right] \qquad (36)$$

For any given ΔE value, the complete reciprocal expression of eqn(36) is also a constant. Extensive functional computer tables of this expression for both monovalent and divalent ions are available[56] for easy C_0 calculations. A small selection is given in Table 13 which assumes the respective slopes to be Nernstian, and takes T at 25°C.

The following is a typical example. A copper solution (100 ml) showed a 4 mV increase after adding 0·1 M copper nitrate solution (1 ml). The known increment C_Δ is readily found from equation 33.

$$C_\Delta = \frac{1 \times 10^{-1}}{100} = 10^{-3} \text{ M}$$

TABLE 13

Some Calculated Values for the Antilog Expressions of Eqns(36) and (39), that is, C_0/C_Δ for Monovalent and Divalent Ions at 25°C.

Potential Change (mV)		Addition Method	Subtraction Method
$\Delta E(M^+$ or $M^-)$	$\Delta E(M^{2+}$ or $M^{2-})$	$\left[\dfrac{1}{(\text{antilog } \Delta E/S)-1}\right]$ $= \dfrac{C_0}{C_\Delta}$	$\left[\dfrac{1}{(1/\text{antilog } \Delta E/S)-1}\right]$ $= \dfrac{C_0}{C_\Delta}$
4	2	5·95	6·95
6	3	3·80	4·80
8	4	2·74	3·74
10	5	2·10	3·10
14	7	1·38	2·38
20	10	0·85	1·85
30	15	0·45	1·45

This addition corresponds to a change in $[Cu^{2+}]$ of 10^{-3} M.

For any divalent ion, the C_0/C_Δ ratio for a ΔE value of 4 mV is 2·74 (Table 13). The original total concentration, C_0, of copper in the 100 ml sample is thus:

$$C_0 \text{ (copper)} = \frac{C_0}{C_\Delta} \cdot C_\Delta = 2\cdot74 \times 10^{-3} \text{ M}$$

The success of known addition depends on certain fundamental assumptions:[54–56]

(i) The volume change of sample following addition is negligible (less than 1 per cent for the above example).

(ii) The accompanying change in total ionic strength is also negligible.

(iii) The fraction of ions actually being measured, that is, uncomplexed, remains essentially constant in the two solutions. Of course, the subsequent arithmetic gives the total amount present, that is, complexed + uncomplexed species.

(iv) Serious interferants are not present in such quantities as to adversely affect the electrode.

(v) The slope is Nernstian. If not, the electrode can be calibrated and the experimentally determined slope substituted for S.

A variation on the known addition technique is to add a small increment of the unknown analytical solution to a larger volume of a known standard. This method is particularly useful when,

(i) only small quantities of sample solutions are available;

(ii) contamination or dilution of sample solutions has to be avoided.

Known Subtraction Method

This is really addition, except that the additive lowers the sample activity (and concentration) by complexation. Some standards, such as sulphide, are notoriously difficult to store at a maintained concentration owing to oxidation. It is in such cases that known subtraction is a particularly useful technique. An additional assumption is made, namely, that the complexing, or precipitating, agent is a much stronger complexing agent than any others present in the original sample. The added reagent must also react completely with the ion to be measured.

Once again, the potential, E_1, of the sample is measured. Sufficient complexing agent is then added, thereby lowering the level of free sample ions, and the potential, E_2, measured. Calculations along lines similar to those for known addition gives an expression,

$$(E_2 - E_1) = \Delta E = S \log \left(\frac{C_0 - C_\Delta}{C_0} \right) \qquad (37)$$

However, $(E_2 - E_1)$ will be negative in known subtraction since $E_2 < E_1$ and, therefore,

$$\frac{-\Delta E}{S} = \log \left(\frac{C_0 - C_\Delta}{C_0} \right) \qquad (38)$$

55

Rearrangement of eqn(38) gives,

$$\frac{C_0}{C_\Delta} = \left[\frac{1}{(1/\text{antilog}\,\Delta E/S)-1}\right] \tag{39}$$

Calculations of C_0 from Known Subtraction Data
Either direct substitution in eqn(38), or the Orion 407 Specific Ion Meter, can be used as with the known addition technique. Tabulated data based on eqn(39) are also extensively available,[56] but a little care is necessary owing to the stoichiometry of complexation. For 1:1 complexation (or precipitation) there is no problem, and eqn(33) is still valid. However, when the added material is divalent, and the sample ion is monovalent, then the appropriate modification for C_Δ is,

$$C_\Delta = \frac{2V_A C_s}{V_0} \tag{40}$$

As an illustration, consider a sulphide solution (100 ml) where the potential, -841 mV, changes to -836 mV after adding 0·1 M silver nitrate (1 ml). Because each sulphide anion reacts with two silver cations, then the equation for C_Δ becomes,

$$C_\Delta = \frac{V_A C_s}{2V_0} = \frac{1 \times 10^{-1}}{2 \times 100} = 5 \times 10^{-4}\ \text{M}$$

This corresponds to a fall in $[S^{2-}]$ of 5×10^{-4} M and not of 10^{-3} M sulphide as in 1:1 stoichiometry.

For a 5 mV potential change, the C_0/C_Δ value is 3·10 (Table 13). Thus C_0 (sulphide)

$$= \frac{C_0}{C_\Delta} \cdot C_\Delta = 3·10 \times 5 \times 10^{-4} = 1·55 \times 10^{-3}\ \text{M}.$$

Sulphide could, of course, be determined by known addition based on sulphide, except that silver is much more convenient to handle than sulphide.

These total concentration methods are also useful in the determination of equilibrium constants.

5

NON-GLASS SELECTIVE ION SENSITIVE ELECTRODES: CLASS A. HOMOGENEOUS SOLID STATE ELECTRODES

All selective ion-sensitive electrodes function on a broadly similar basis. However, there are various mechanisms whereby the primary and interferant ions are either allowed, or forbidden, access to move across the all-important membrane.

The electrodes fall conveniently into three classes, namely, homogeneous solid state, heterogeneous solid state and liquid ion exchanger membrane electrodes. Little historical attention is paid to the early attempts to fabricate these selective electrodes. Some examples of each class will be described, together with a selection of their applications.

The principal advantages of solid state homogeneous electrodes lie in their low cost, low response times, long operative life, resistance to corrosive acid and alkaline media, Nernstian behaviour over many activity decades, and the freedom from redox interferences such as permanganate. The important specification parameters, and the addresses of companies manufacturing solid state electrodes are given in Tables 14 to 17. These are essentially manufacturers listings and may suffer periodic alteration; for example, the Orion 94-58 thiocyanate electrode has been withdrawn.[57] Manufacturers' details may also be at variance with independent research publications. The pH ranges are often listed without regard to concentration dependence. Selectivities are listed separately (Tables 7 and 8). Patents[58–62] contain a wealth of technical details.

57

TABLE 14

Beckman Solid State Membrane Electrodes

Addresses: 2500, Harbor Boulevard, Fullerton, Calif., 92634, U.S.A. (P.O. Box 1, Glenrothes, Fife, Scotland).

Ion	Model	Molar Range	pH Range	Resistance (Megohms)	Response Time (Seconds)*	Principal Interferants
F^-	39600	$10^\circ \rightarrow 10^{-6}$	$0 \rightarrow 13$†	$\vee\ 5$	$\vee\ 3$	OH^-
Cl^-	39604	$10^\circ \rightarrow 5 \times 10^{-5}$	$0 \rightarrow 14$	$\vee\ 1$	$\vee\ 2$	Br^-; I^-; S^{2-} and CN^-
Br^-	39602	$10^\circ \rightarrow 10^{-7}$	$0 \rightarrow 14$	$\vee\ 1$	$\vee\ 2$	I^-; S^{2-} and CN^-
I^-	39606	$10^\circ \rightarrow 10^{-8}$	$0 \rightarrow 14$	$\vee\ 0.25$	$\vee\ 3$	S^{2-}
S^{2-}	39610	$10^\circ \rightarrow 10^{-9}$	$0 \rightarrow 14$	$\vee\ 0.25$	$\vee\ 3$	
Cu^{2+}	39612‡	10^{-8} (lower limit)				Ag^+ and Hg^{2+}

Operative temperature range (°C)   : $-5 \rightarrow 100$.
Overall size, length × diameter, (cm):12·8 × 1·25.
* Defined as the time required to obtain a 90 per cent response to a stepchange from $10^{-4} \rightarrow 10^{-3}$ M concentration in a stirred solution. Will also depend on concentration and viscosity of samples. For the copper electrode up to several minutes may be required.
† At 10^{-1} M fluoride.
‡ Like the Coleman solid state electrodes contains no internal filling solution.

TABLE 15

Coleman Solid State Membrane Electrodes*

Addresses: 42, Madison Street, Maywood, Illinois 60153, U.S.A. (Perkin-Elmer Ltd., Beaconsfield, Bucks, U.K.).

Ion	Model Number	Molar Range	pH Range	Resistance (Megohms)	Membrane Material	Principal Interferants
Br^-	3–801	$1 \rightarrow 10^{-7}$	$0 \rightarrow 14$	< 10	AgBr	S^{2-} ; CN^-
Cl^-	3–802	$1 \rightarrow 10^{-6}$	$0 \rightarrow 14$	< 10	AgCl	S^{2-} ; CN^-
F^-	3–803	$1 \rightarrow 10^{-6}$	$4 \rightarrow 8†$	< 0.5	LaF_3	OH^-
Cu^{2+}	3–804	$1 \rightarrow 10^{-6}$	$0 \rightarrow 14$	< 100	CuS	Ag^+ ; Hg^{2+}
S^{2-}		$1 \rightarrow 10^{-7}$			Ag_2S	
Ag^+	3–805	$1 \rightarrow 10^{-7}$	$0 \rightarrow 14$	< 0.05		Hg^{2+}

Operative temperature range (°C): $-5 \rightarrow 100$.
Time response: decreased by stirring.
Overall size, length × diameter (cm): 14.3×0.8.
Dollar Cost: 150 introductory offer.

* Contain no inner solution and can be used upsidedown without modification for micro assay of droplets.
† pH $0 \rightarrow 13$ under particular specified conditions.

TABLE 16
Orion 94 and 96 Series Solid State Specific Ion Electrodes
Addresses: 11 Blackstone St., Cambridge, Mass., 02139, U.S.A. (E.I.L., Richmond, Surrey)

Ion	Model	Total Molar Concentration Range*	pH Range†	Resistance (Megohms)	Membrane Material	Principal Interferant Ions
CNS^-	94-58	$10^0 \rightarrow 10^{-5}$	$0 \rightarrow 14$		$AgSCN + Ag_2S$	S^{2-}, Hg^{2+} and Cu^{2+} must be absent
CN^-	94-06	$10^{-2} \rightarrow 10^{-6}$	$3 \rightarrow 14$	< 30	$AgI + Ag_2S$	S^{2-} must be absent
F^-/La^{3+}	94-09	$10^0 \rightarrow 10^{-6}$	$0 \rightarrow 11$	< 1	$LaF_3 + Eu^{\ddagger}$	} OH^- only interference
F^-/La^{3+}	96-09	$10^0 \rightarrow 10^{-6}$	$0 \rightarrow 11$	< 30	$LaF_3 + Eu^{\ddagger}$	
Na^+	94-11	$10^0 \rightarrow 10^{-6}$	$3 \rightarrow 12$	< 200		Ag^+
Cl^-	94-17	$10^0 \rightarrow 5 \times 10^{-5}$	$0 \rightarrow 13$	< 30	$AgCl + Ag_2S$	} S^{2-} must be absent
Cl^-	96-17	$10^0 \rightarrow 5 \times 10^{-5}$	$0 \rightarrow 13$	< 30	$AgCl + Ag_2S$	
Br^-	94-35	$10^0 \rightarrow 5 \times 10^{-6}$	$0 \rightarrow 14$	< 10	$AgBr + Ag_2S$	S^{2-} must be absent
I^-	94-53	$10^0 \rightarrow 5 \times 10^{-8}$	$0 \rightarrow 14$	$1 \rightarrow 5$	$AgI + Ag_2S$	S^{2-} must be absent
S^{2-}	94-16 }	$10^0 \rightarrow 10^{-7}$	$0 \rightarrow 14$		Ag_2S	None as far as examined
Ag^+		$10^0 \rightarrow 10^{-7}$	$0 \rightarrow 14$	< 1	Ag_2S	Hg^{2+} must be absent
Cu^{2+}	94-29	$10^0 \rightarrow 10^{-7}$	$0 \rightarrow 14$	< 1	$Ag_2S + CuS$	S^{2-}, Ag^+ and Hg^{2+} must be absent
Cd^{2+}	94-48	$10^0 \rightarrow 10^{-7}$	$1 \rightarrow 14$	< 1	$Ag_2S + CdS$	Ag^+, Hg^{2+} and Cu^{2+} must be absent
Pb^{2+}	94-82	$10^0 \rightarrow 10^{-7}$	$2 \rightarrow 14$	< 1	$Ag_2S + PbS$	Ag^+, Hg^{2+} and Cu^{2+} must be absent

Operative temperature range (°C) : About $0 \rightarrow 80$ continuous usage; $0 \rightarrow 100$ intermittent usage.
Overall size, length $\times$ diameter (cm):13.9×1.2; 13.9×1.3 for combination electrodes (slightly larger dimensions were once quoted).
Minimum sample size (ml) : 5 in 50 ml beaker ($10 \rightarrow 20$ for 94–06 Model); 0.5 in disposable Orion micro sample dish. Combination electrodes can take 10 microlitres using spot test paper.
Reproducibility : Drift, repeatability and response time characteristics are generally comparable with those of a good quality pH electrode.
Storage : Can be dry stored or in appropriate ion solution. Pre-soaking is unnecessary. Model 94–

Dollar cost : Ranges from 150 → 250. Combination electrodes (96 series) about $40 more than analogous 94 series electrodes.

* Refers to total concentration (complexed and uncomplexed ions).
† Range for 10^{-3} M activity level of measured ion.
‡ Europium (II) is said to improve conductivity but is not essential for electrode functioning.

TABLE 17

Philips Solid State Electrodes

Addresses: Eindhoven, Netherlands (Pye Unicam, York St., Cambridge, U.K.)

Ion	Model Number	Molar Range	pH Range*	Resistance (Megohms)	Membrane Material†	Principal Interferants‡
Cl⁻	IS 550–Cl	$10^{-1} \rightarrow 5 \times 10^{-5}$	$0 \rightarrow 14$	1	AgCl	S^{2-}; Ag^+; NH_3 ($CN^- > I > S_2O_3^{2-} > 1:2$)‖
Br⁻	IS 550–Br	$10^{-1} \rightarrow 10^{-6}$	$0 \rightarrow 14$	1	AgBr	S^{2-}; Ag^+; NH_3 ($CN^- > I > S_2O_3^{2-}$)‖
I⁻ ⎫	IS 550–I	$10^{-1} \rightarrow 10^{-6}$	$0 \rightarrow 14$¶	1	AgI	S^{2-}; Ag^+; NH_3
CN⁻ ⎭	or IS 550–CN					(I^-)‖

Operative temperature range (°C) : 90 → 100
Cost : £50
Storage : Can be air stored. A one-hour presoaking in 10^{-1} M relevant primary ion solution is recommended before use.
Response times(s) : 20.
Reproducibility : ± 2 mV.
Electrode life : > 12 months.
  * : For 10^{-1} M solution of relevant anion.
  † : Replaceable membranes provided.
  ‡ : Electrodes show no response to cations except Ag^+.
  ‖ : Parenthesized ions should be interferants since selectivities > 1 (Table 8) but are not specifically listed by Philips as are S^{2-}, Ag^+ and NH_3.
  ¶ : Should make pH > 11 with NaOH for use in CN^- media.

Solid state electrodes are not new. Promising attempts were made over thirty years ago to fabricate electrodes from silver halide discs,[63] slices of barium sulphate,[64] and calcium fluoride.[64] Electrodes incorporating these materials never enjoyed wide analytical usage, owing to the rather high electrical resistance of silver chloride, for example, and the serious photoelectric effects of all silver halides studied,[63] as well as lack of commercial availability.

There are certain stringent requirements which any material must satisfy before it can act as a successful membrane in a solid state electrode device. It must:

(i) be substantially non-porous, that is, imporous;

(ii) solicit the minimum photoelectric response;

(iii) have good mechanical strength and not be readily scratched or abrazed;

(iv) be available as a large crystal, or else easily made by pressing. During the pressure process, stress may cause decomposition to metal. This metal can be a serious drawback, since the electrode could then be subject to interferences from redox systems;

(v) have medium bulk resistivities;

(vi) be water insoluble, in fact highly insoluble; the smaller its solubility product, K_s, the better. The solubility should be such that any ion activity arising directly in the sample solution from the membrane is less than that lowest activity likely to be faced in any analytical situation;

(vii) show good selectivity. As already discussed (p. 31) this is directly related to the solubility product of the membrane material, and the solubility product of that compound which may, or may not, form by reaction of the interferant ion and an oppositely charged ion constituting the membrane;

(viii) have good time response characteristics;

(ix) be capable of easy tight sealing to the electrode body, otherwise low impedance leak paths arise between the internal reference solution and sample solution;

(x) if possible, exhibit Nernstian response.

It is largely through the pioneering efforts of Frant and Ross during the mid sixties that materials matching these requirements have been successfully fabricated and translated into practical and commercial reality (Table 16).

Lanthanum fluoride, silver chloride, bromide and iodide are crystalline materials in which the ionic conduction process at room temperature is attributed either to the anion or the cation, usually depending on which has the lower charge and smaller ionic radius. Both factors are ideally suited to the lattice defect conduction mechanism, in which the mobile ion moves into an adjacent lattice defect position. Other ions of different dimensions and charge are unable to move in this fashion, and cannot participate in the conduction. Interference is not *usually* caused by the foreign ions entering the crystal lattice, but by chemical reactions at its surface (p. 28). Selectivity is due to this entry restriction of all ions except the one of analytical concern. However, if M and N are of the same size and charge, for example, F^- and OH^-, then interference can arise by competitive conduction.

In other cases, for example, the solid state halide electrodes comprising Ag_2S-AgX compounds, an ionic-electronic conduction mechanism is involved.[65]

The Lanthanum Fluoride Electrode

Construction and Response Characteristics
One of the few truly specific ion sensitive electrodes was first introduced by Ross and Frant.[66] Briefly, this remarkable fluoride electrode consists of a hollow, rigid, polyvinyl-chloride or polytetrafluorethylene tube which is electrically insulating and chemically inert to sodium hydroxide (10 per cent), nitric acid (25 per cent), hydrofluoric acid (10 per cent), methanol, ethanol, benzene, acetonitrile and glacial acetic acid. One end is completely sealed with a lanthanum fluoride crystal membrane (diameter 1 cm and 1–2 mm thick). An internal Ag-AgCl reference electrode, immersed in the

63

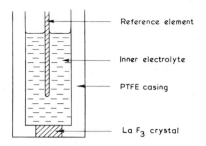

Reference element

Inner electrolyte

PTFE casing

La F$_3$ crystal

Fig. 15. Section through fluoride crystal membrane electrode.

chloride-fluoride internal filling solution, is then screwed in at the other end of the stem (Fig. 15).

In the early models, the crystal membrane was held in place by pressure wedge-sealing, but since then all Orion solid state crystal membranes have been fabricated by an epoxy resin sealing technique. Electrodes thus sealed give better performances from the standpoint of preventing internal solution leakage. Also, the external membrane surface is flush with the polymer body and thus prevents occlusion.[58] This prevents residues of solutions occluded in these zones being carried over into fresh analytical samples, a factor causing spurious results, or electrode drifting, which only settled when the "two solutions" became mixed.[58]

The e.m.f., E, of the cell

Ag; AgCl(s) $\left|\begin{matrix} F^-(0\cdot1\ \text{м}) \\ Cl^-(0\cdot1\ \text{м}) \end{matrix}\right.$ $\left|\begin{matrix} LaF_3(s) \\ EuF_2(s) \end{matrix}\right.$ $\left|\begin{matrix} \text{Test} \\ \text{Sample} \end{matrix}\right|$ Reference Electrode

$\longleftarrow E_{R'} \longrightarrow$ $\longleftarrow E_{M(int.)} \longrightarrow$ $\longleftarrow E_{M(ext.)} \longrightarrow \longleftarrow E_j + E_{R''} \longrightarrow$

$\longleftarrow$————Fluoride Electrode———$\longrightarrow$

is given by:

$$E = E_{R'} + E_{M(int.)} + E_{M(ext.)} + E_j + E_{R''} \qquad (41)$$

64

A constant potential, $E_{R'}$, is developed between the Ag/AgCl internal electrode and the internal fluoride-chloride solution of fixed concentration, and similarly another constant potential, $E_{M(int.)}$, exists between this same halide filling solution and the inner surface of the lanthanum fluoride membrane. Eqn(41) can thus be considerably simplified to,

$$E = E^\circ - 2 \cdot 303 \frac{RT}{F} \log a_{F^-} \qquad (42)$$

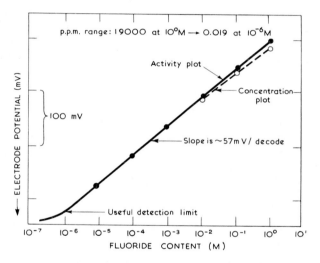

Fig. 16. A typical calibration plot of an Orion 92-09 fluoride electrode (as plotted on semi log paper).

where E is the total measured potential, and E° is the algebraic sum of the potentials of the internal Ag/AgCl electrode, $E_{R'}$, the reference electrode potential $E_{R''}$, the liquid junction potential, E_j, and the potential established at internal membrane surface, $E_{M(int.)}$. The actual value of E° for the fluoride electrode, or any selective ion sensitive electrode, will depend

65

on the choice of internal and external reference electrodes, as well as the internal filling solutions and, of course, the magnitude of the liquid junction potential. The potential taken in a fluoride solution of unit activity at 25°C is obviously the constant $E°$, but this is more commonly measured from the results of an electrode calibration.

A typical e.m.f./activity calibration is shown in Fig. 16. Below 10^{-6} M fluoride, the electrode begins to fail owing to the finite solubility of the lanthanum fluoride membrane. The solubility product of this sensor material—here as a massive crystal—is, therefore, about $(10^{-6})^3 \times (3\cdot3 \times 10^{-7})$, that is, $3\cdot3 \times 10^{-25}$, although Frant and Ross[66] quote about 10^{-29} based on a lower fluoride level.

It is, therefore, possible to calculate, at least in principle, the approximate detection limit of any solid state electrode from solubility product data. However, some care is needed since both the above values are at variance with that of 7×10^{-17} obtained from a potentiometric titration of fluoride against lanthanum nitrate;[72] this, if correct, would set the theoretical detection limit much higher at around 10^{-4} M. This hundred-fold discrepancy could be related to the slow fluoride membrane dissolution time, or a big difference between the solubility product of the solid crystal membrane and freshly precipitated lanthanum fluoride. Later, Lingane published a slightly smaller value, about $1\cdot2 \times 10^{-18}$, and further maintained this to be nearly, or perhaps exactly, the same for both forms of material.[73]

Whatever the value may be, there is wide and general agreement on the 10^{-6} M detection limit for all three commercial fluoride electrodes (Tables 14, 15 and 16), but this limit is restricted by pH. Unfortunately, a closer comparison in most other respects (particularly for their applications) is impossible at this time owing to the limited primary journal and technical information on Beckman and Coleman solid state fluoride electrodes. However, there seems to be no good reason for preferential use of any one fluoride electrode, although the Orion 96-09 combination model has some obvious practical advantages. Similar considerations possibly apply to any other particular electrode type marketed by more than one manufacturer.

Response times of these fluoride electrodes are variously reported to lie in the few seconds to minutes range. The lower the fluoride activity the longer the response time, which can stretch to around one hour for 10^{-6} M solutions provided the pH is in the right range.

The Orion electrode will withstand mechanical and thermal shocks without cracking or breaking. Furthermore, no soaking or conditioning is necessary before use, and the durable membrane is not poisoned by proteins or other organic materials.

Interferences

Frant and Ross[66] found the potential recorded in 10^{-4} M fluoride solution decreased by about 7 mV after adding 10^{-1} M chloride, or nitrate, and by about 20 mV after adding 1 M chloride or nitrate. These values are within 10 per cent of those expected on the basis of the accompanying increased ionic strength of the mixed solutions. This activity interference is of the type already briefly described (p. 32). The fluoride electrode does not respond to these anions, or to bromide, iodide, sulphate, bicarbonate, phosphate or acetate anions, even when present in a thousand-fold excess. The only important interference is the hydroxide ion, which becomes significant when $[OH^-] \geq [F^-]$. A ten-fold excess of hydroxide will double the apparent fluoride level.[66] The typical potential behaviour of the Orion 94-09 electrode as a function of three fluoride activities and variable pH is shown in Fig. 17. The hydroxyl interference is attributed[66] to their similarities in size and charge. While this mechanism may be partly operative, an equilibrium of the type,

$$LaF_3(s) + 3OH^- \rightleftharpoons La(OH)_3(s) + 3F^- \qquad (43)$$

seems more realistic since the fluoride released raises the

sample fluoride level to give a more negative potential—as indeed observed (Fig. 17). However, the mechanism of hydroxyl interference is probably more complicated owing to various other lanthanum complexation equilibria. The extent of interference, irrespective of low or high pH, must be related to the fluoride activity or concentration, and the lower the fluoride level the more restricted is the pH range. Indeed in 10^{-6} M fluoride, the effective pH range is very narrow, and some confusion may arise over the term "pH range" on two counts. Firstly, manufacturers' manuals quote three different pH ranges ($1 \rightarrow 8$, $0 \rightarrow 7$ and $0 \rightarrow 8.5$), for the electrode in 10^{-6} M fluoride. Secondly, around pH $4 \rightarrow 5$ the formation of HF and HF_2^- begins to lower the fluoride activity, thereby giving more positive potential readings. This is an excellent example of the other type of interference mentioned already, namely, that due to complexation (p. 32). However, even below pH 5, the electrode is still responding correctly to the fluoride, that remains uncomplexed, and as shown in some manufacturers' diagrams, notably for the Beckman electrode, with a near Nernstian slope of about 60 mV per decade.

In effect, a pH range, when quoted numerically, should:
(i) exclude that section of the plot which responds to interfering ions for example, at high pH in this case;
(ii) be accompanied by the concentration, or activity, of the primary ion involved.

Thus, it is quite realistic to quote a proper pH range of $0 \rightarrow 11$ for 10^{-1} M fluoride on the basis of Fig. 17, rather than $5 \rightarrow 11$, since the electrode is still functioning faithfully to fluoride activity below pH 5. A pH range of $0 \rightarrow 14$ for any electrode would mean no pH interference at one or more particular primary ion activities, and the corresponding e.m.f.—pH plots would comprise a series of horizontal lines. Considerable care is thus necessary in the interpretation of pH ranges listed in some of the manufacturers' specification tables.

Even detrimental pH interference conditions can usually

68

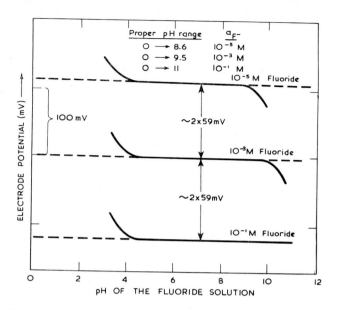

Fig. 17. *Typical effect of pH on the potential response of an Orion fluoride electrode at three different fluoride activities.*

be overcome by adding acid or alkali to bring the sample pH within the middle horizontal section of an e.m.f.—pH plot.

Other Fluoride Membrane Materials

The quantity of europium(II) fluoride used to dope the original lanthanum fluoride membrane remains proprietory information.[66] Europium (II) is extremely susceptible to aerial oxidation and europium(III) must be formed, at least in the membrane surface.[73] Europium(III) fluoride (K_s 2·2 × 10^{-17}) is more soluble than lanthanum fluoride, and if present in considerable quantity, should lower the fluoride sensitivity of the electrode. The doping is said to improve the crystal conductivity, but the necessity for this is uncertain.

69

Lingane concludes,[73] with good reason, that the quantity of europium(II) dope in the Orion electrode is likely to be very small.

Other crystalline materials suitable for functional membranes include[61,66] cerium, praseodymium and neodymium trifluorides. The selectivity of mixed lanthanide membranes, for example, LaF_3-NdF_3, is as good as that of single salt membranes, but the electrodes tend to be noisier owing to higher bulk resistivities.[61] Scandium, yttrium, lead and bismuth fluorides are also quoted as suitable materials,[61] but growing large crystals of bismuth trifluoride is difficult. However, compression of bismuth trifluoride at 5×10^4 p.s.i. and $500 \rightarrow 550°C$ gives membrane material with good fluoride response, but chloride and sulphate present serious interference problems for useful work in hard waters.[61]

Other possible materials, such as calcium fluoride, are insufficiently sensitive, since the solubility product, $K_s = 3.9 \times 10^{-11}$, would in principle generate about 4×10^{-4} M or 8 p.p.m. of fluoride.

Applications of the Lanthanum Fluoride Electrode
Since 1966, the Orion electrode has had a considerable run of analytical successes.[67] About eighty papers have been published, almost exceeding the number of papers dealing with all other selective ion sensitive electrodes. These studies cover water,[61,68,69] beverages,[70] milk,[71] potentiometric titrations,[72,73] complexation,[74] reaction rates,[75] urine,[76] saliva,[77] serum,[78] toothpaste,[79] bone,[80] mineral fluorides,[81,82] organic compounds,[83] air and stack gases[84] and nuclear fuel processing solutions.[85] Sample pretreatment is, of course, necessary for rocks and bones in order to bring the fluoride into solution.

Whatever the relative merits of fluoridation of potable water supplies, the concentration, rather than activity, of the fluoride in such waters is important for quality control. Problems arise in such analyses owing to the wide variation of ionic strengths of samples as well as the possible presence

70

TABLE 18

The Composition and Ionic Strength of T.I.S.A.B. pH ~ 5[68]

Composition	z^2C
Sodium Chloride (1M)	2
Acetic Acid (0·25 M)	0
Sodium Acetate (0·75 M)	1·5
Sodium Citrate (0·001 M)	Negligible

$$\mu = \tfrac{1}{2}\Sigma z^2 C = 1\cdot 75 \text{ M}$$

of complexing cations (Fe^{3+} and Al^{3+}). An elegant and rapid method based on a 50:50 dilution of both samples and standards with Total Ionic Strength Adjustment Buffer (T.I.S.A.B.) seems to have largely solved these difficulties.[68]

This high ionic strength treatment fulfils three simultaneous rôles:

(i) The total ionic strength of the water samples is essentially fixed since the high level of T.I.S.A.B. ions swamps the variation in the original water. Furthermore, the T.I.S.A.B. reagent contains no interfering ions.*

---

* Anfält and Jagner (*Anal. Chim. Acta* (1969), **47**, 483), have stated that "carboxylic acid buffer systems are unsuitable for all work with this (fluoride) electrode". Their preliminary work indicates the electrode to be sluggish in response, and to have non-Nernstian characteristics in acetate media. These observations mitigate against the use of T.I.S.A.B., but, nevertheless, the latest Orion T.I.S.A.B. solution (Orion Research Inc., Applications Bulletin No. 5A, 1969) retains acetate, although C.D.T.A. (1,2-diaminocyclohexane-N,N,N′,N′-tetraacetic acid) is now used instead of citrate when analytical samples contain aluminium or iron. However, acetate, citrate, formate, oxalate do not present any serious interferences (Evans, Moody and Thomas, unpublished work).

(ii) The samples are buffered by the acetic acid-sodium acetate in the optimum range of $5 \rightarrow 5.5$, thereby avoiding the only known serious interferant (hydroxyl) for the fluoride electrode.

(iii) The citrate reacts with any iron or aluminium fluoro-complexes and liberates the bound fluoride in the detectable free form.

Even before this improved technique became available, between 50 and 100 routine water analyses could be performed each day.[86]

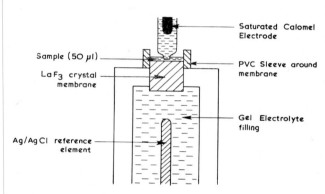

Fig. 18. The Durst-Taylor modification of the Orion 92-09 fluoride electrode (Ref. 87).

Modifications of Solid State Electrodes in General, and the Lanthanum Fluoride Electrode in Particular

Even with a special microdish facility, sample size must still be at least 0.5 ml, but an ingeniously modified Orion fluoride electrode takes 50 μl, and even 25 μl, samples.[87] The inner electrolyte solution is drained and replaced by a hot gel (0.1 M KCl + 0.1 M NaF in 4 per cent agar) which sets, thereby making an "immovable" contact between the inner membrane surface and electrolyte, and the Ag/AgCl reference electrode and inner electrolyte. A pvc sleeve forms a tight seal around the protruding membrane, and on inversion

72

becomes the sample container into which is dipped a calomel reference electrode (Fig. 18). The general performance of the modified electrode compares favourably with that of the "normal" electrode, except that the Nernstian range stops around 3×10^{-5} M although useable response is claimed[87] to 5×10^{-6} M.

The "normal" electrode cannot function on inversion because the aqueous inner electrolyte would drop below the inner membrane surface. Also, a different means of terminal containerization would be needed with present-day flush membrane models (Fig. 15).

Inversion is not a problem with some electrodes; all the solid-state Coleman electrodes, the Beckman copper electrode, and at least the Orion sulphide electrode no longer contain an inner electrolyte solution. Instead, the inner reference electrode makes a direct contact seal with the inner surface of the sensor membrane material.

Another most important advance is the combination electrode[60] (compare with the glass combination electrode, Fig. 12). So far, only the Orion fluoride 96-09 and chloride 96-17 models are available. Their overall performance is comparable with the corresponding Orion 94 series models. With these sample volumes, as little as 10 μl can be assayed by first absorbing into a small test paper lying on a microscope slide. The sensor end of the combination electrode, whose area matches the paper, is then firmly pressed into the moistened paper. A stable potential can be obtained in about 30 seconds.

These modified systems represent a miniaturization of sample size, and not of electrodes; the latter are quite bulky, although the combination electrodes are slightly smaller than their single counterparts (Table 16). Apart from some work on calcium liquid ion exchanger electrodes,[6,88] there is no ready information on the true miniaturization of selective ion sensitive electrodes.

Solid state electrodes with a flow-through configuration are not available as standard production items, but in certain cases they may be made to order.[57]

73

Silver Halide Electrodes

The difficulties associated with the adoption of pure silver halides for use as electrodes have already been outlined (p. 62). However, a technique[59] incorporating silver sulphide, which is itself a chemically inert matrix, provides highly suitable electrode materials. An intimate mixture of,

TABLE 19

Characteristics of Silver Chloride-Silver Sulphide Discs*
Pressed between 10^4 and 2.5×10^4 p.s.i.[59]

Sample Number	Mole % AgCl:Ag$_2$S	Resistivity (ohm cm^{-1})	Photo-electric Effect (mV)†	General Character	Nernstian Response in Sodium Chloride
1	100:0	10^7	> 30	Glassy, clear, soft‡	to 10^{-5}M
2	90:10	10^4	< 1.5	Black, shiny, hard, dense, imporous and scratch resistant	to 10^{-5}M
3	50:50	7×10^4	~ 3	Harder and stronger than sample 1, but not so hard or strong as sample 2	to 10^{-5}M
4	10:90	1.5×10^4	< 1.5	Hard, dense and as scratch resistant as sample 2	to 10^{-3}M
5	0:100	< 10^5‖	< 1.5	—	nil response

* About 0.4 inch diameter and of 0.06 inch thickness.

† Represents potential change when the ambient illumination varies from dark to typically ordinary room lighting conditions.

‡ Can be scratched with the fingernail.

‖ See reference 23.

say, Ag$_2$S-AgCl, is easily obtained by adding a two-fold stoichiometric excess of silver nitrate to a solution of sodium sulphide nonahydrate and sodium chloride of the apropriate molarity.[59] After at least twenty successive water washings to remove nitrate traces, the precipitate is dried around 110°C and is ready for disc pressing (Table 19).

Similar trends are observed with silver bromide, except that photoelectric effects are insignificant. Silver iodide pressed discs are fragile and tended to break on removing from the die-press.[59] While $AgCl$-Ag_2S discs are believed to be true mixtures, the corresponding bromide and iodide materials might comprise substantial proportions of the compounds Ag_3SBr and Ag_3SI formed during the pressing stages.[59]

Such mixed, pressed, discs incorporated into Orion electrodes meet all the stiff requirements demanded of solid state membranes (p. 62). After epoxy sealing a silver chloride-silver sulphide disc into a fluorocarbon body (re the lanthanum trifluoride electrode, Fig. 15) containing an internal silver-silver chloride electrode dipping into the necessary internal halide electrolyte, the resulting electrode gives excellent analytical performances.[59] Coleman halide electrodes contain no internal filling solution. The Philips and Coleman electrodes are interesting in that pure silver halides are used without the apparent silver sulphide inclusion for the sensor membrane.

Although fewer technical details are available for other halide electrodes, their general specification and selectivities compare favourably (Tables 7, 8 and 14–17).

All the potentials, except the potential at the external membrane—sample interface, are fixed, and the potential E of the cell,

Ag,AgCl(s) | 0·1 M Cl⁻ |Ag$_2$S — AgCl(s) |Test Sample | Reference Electrode

is given by,

$$E = E^\circ + 2\cdot303 \frac{RT}{F} \log a_{Ag^+} \qquad (44)$$

Again, as with the fluoride electrode, the constant term, E°, is the sum of all the constants of the system, but is, of course, temperature dependent. A certain silver activity, albeit small, will arise in the sample solution, even if originally devoid of silver, owing to the finite solubility of silver chloride. The silver halide is in equilibrium with the sample, and the resulting silver activity is given by,

$$a_{Ag^+} = K_{s_{AgCl}}/a_{Cl^-} \qquad (45)$$

Substituting this silver activity into eqn(44), gives

$$E = E^\circ + 2 \cdot 303 \frac{RT}{F} \log \frac{K_{s_{AgCl}}}{a_{Cl^-}} \qquad (46)$$

Provided the temperature remains constant, then $K_{s_{AgCl}}$ is also constant, and,

$$E = E^{\circ\prime} - 2 \cdot 303 \frac{RT}{F} \log a_{Cl^-} \qquad (47)$$

The new constant, $E^{\circ\prime}$ is given by,

$$E^{\circ\prime} = E^\circ + 2 \cdot 303 \frac{RT}{F} \log K_{s_{AgCl}} \qquad (48)$$

This means that the mixed membrane behaves just like a pure silver chloride ionic conductor, in which the silver cation (1.26 Å radius), and not the larger chloride anion (1·81 Å radius), is the mobile species. Indeed, the calibration for the mixed silver chloride-silver sulphide electrode obeys eqn(47) over a fairly wide chloride activity range.

In theory, similar suitably fabricated membranes can be made (and obeying equations analogous to equation 47) for any anion M^-, provided that:

(i) $K_{s_{AgM}} < K_{s_{Ag_2S}}$ (unlikely since $K_{s_{Ag_2S}} \sim 10^{-51}$);

(ii) $K_{s_{AgM}}$ is not too large and thereby give a high M^- solution activity in its own right.

The detection limits of electrodes require definition. Ideally, they should be the break-away point at which the linear portion of the calibration becomes non-linear. However, "limits of detection" are sometimes quoted instead, and these can be even two decades beyond this breakaway point. The theoretical detection limits of halide electrodes calculated from solubility product data (Table 9) are compared, in Fig. 19 with the detection limits (as defined above) for three Beckman silver halide electrodes.[89] The agreement for iodide is excellent but the discrepancy for bromide or chloride (Fig. 19) is also evident in Table 14, whose data is compiled from a Beckman bulletin.[90]

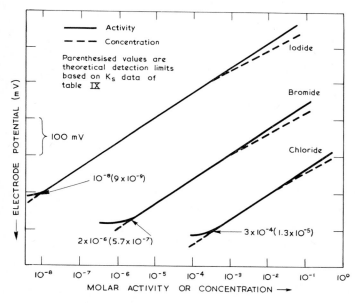

Fig. 19. Calibration plots for three Beckman halide electrodes.

Another characteristic feature of many manufacturers' calibration plots for solid state electrodes (Figs. 2, 3 and 19) is the more gradual onset of interference, which contrasts with the abrupt onset shown in Fig. 11. These differences could well be due to the lack of experimental points in the twilight zones.

Operational Lifetime of Silver Halide Electrodes

Apart from the Orion 94-06 cyanide electrode (with a lifetime of about 4 months in 10^{-5} M solutions, and of several hundred hours in 10^{-3} M solutions), the available literature lists lifetimes which are often greater than one year. One Orion 94-17 chloride electrode has been used continuously for some 20,000 soil analyses.[55] In this instance,[55] the badly grooved, scarred membrane was restored to "as-new" condition with a glossy, unmarked appearance by abrasive polishing within less than two hours. After calibration, the reconditioned chloride electrode gave correct e.m.f. readings and with the expected response times.[55]

Applications of Silver Halide Electrodes

A bibliography[67] lists fifteen papers concerning use of solid-state silver halide electrodes; perhaps the most unusual relate to cystic fibrosis studies.[91,92] Few diseases can be so readily diagnosed as cystic fibrosis, because the sweat of over 98 per cent of C.F. patients contains up to five times as much sodium chloride as normal subjects (< 50 m equiv. per litre). Yet a reliably fast sweat evaluation has been difficult for patient and analyst alike.[93]

A reliable diagnosis is now possible using the Orion 96-17 combination chloride electrode.[93] This can cope with patients who fail to sweat copiously following, for example, localized skin heating, or pilocarpine iontophoresis. The following advantages, relative to previous methods, are claimed[93] for the solid state electrode procedure.

(i) It requires the least amount of sweat (few microlitres). The chloride-reference combination electrode is simply pressed on the skin surface, and the need for sweat collection is eliminated. The response, reached in 20/30 s, can be translated directly into milli equivalents per litre on a special portable skin meter.

(ii) It is the most reliable to date. In one series of tests,[91] also conducted with 47 controls (normal subjects), 73 out of 75 cases of cystic fibrosis patients were confirmed.

78

(iii) The speed (5–6 min total) permits large scale, low cost daily screening. Over forty thousand patients have been screened.[93]

(iv) Its simple operation, particularly with a direct readout meter attachment, greatly eases the training of medical technicians for the three-step procedure: just a sweat inducement period, application of electrode, and finally read-out of sodium chloride.

The Silver Sulphide Electrode

The membrane of the Orion, Coleman and presumably Beckman sulphide electrodes comprises silver sulphide. The dense, non-porous, membrane of considerable insolubility makes for an electrode which,

(i) has a very low detection limit which is related to solubility product;

(ii) is resistant to oxidizing and reducing agents;

(iii) senses either silver or sulphide ions;

(iv) has fast response times.

The e.m.f., E, of the cell,

Ag; AgCl(s) | Ag$^+$ (0·1M) | Ag$_2$S(s) | Test Sample | Reference Electrode
←——Silver Sulphide Electrode——→

is given by

$$E = E^\circ + 2{\cdot}303 \frac{RT}{F} \log a_{Ag^+} \qquad (49)$$

A miniscule amount of silver is generated by the membrane even when test samples contain no silver. The resulting silver ion activity, which is dependent on the sulphide ion activity of any test sample, is calculated from,

$$a_{Ag^+} = \sqrt{K_{S_{Ag_2S}}/a_{S^{2-}}} \qquad (50)$$

Combination of eqns(49) and (50) provides an expression relating the e.m.f. to sulphide sample activity,

$$E = E^{\circ\prime} - 2{\cdot}303 \frac{RT}{2F} \log a_{S^{2-}} \qquad (51)$$

79

The new constant $E^{\circ'}$ incorporates an additional factor involving the solubility product, K_s, of silver sulphide.

A solubility product calculation indicates that the electrode should manage to detect about 1.26×10^{-17} M Ag^+ and 6.3×10^{-18} M S^{2-} respectively. It is difficult to reconcile these figures with statements that 10^{-24} M, and even 10^{-25} M Ag^+ and S^{2-}, levels can be measured. The higher of these figures represents a fraction (0.6) of an ion per litre.

The actual detection limit, as shown by calibration, is much higher, namely $\sim 10^{-7}$ M (Table 16). This is due to the experimental difficulty of providing actual calibrant solutions for Ag^+, or S^{2-} ions owing to contamination during serial dilution handling. This problem, of course, is not peculiar to these ions. However, situations arise which realize these ultramicro Ag^+ or S^{2-} levels. Thus, sulphide levels down to about 10^{-20} M can be traced* during the unusually large titration end point-break with silver nitrate (Fig. 20). This means that the range limit (10^{-7} M) given in Table 16 for the 94-16 Orion sulphide electrode is not strictly true. Indeed, the same remark applies to the cadmium and lead figures (Table 16) which can extend beyond 10^{-10} M in solutions containing 10^{-7} M or more total cadmium or lead. Orion state in fact that a detection range of 10^{-17} M, or less, is possible with the 94-16 sulphide electrode in the presence of 10^{-7} M total sulphide or silver.

Interferences

Mercury and silver sulphides are the least soluble compounds known. For this very reason, there can be no real interferences when the silver sulphide electrode is functioning in a sulphide capacity. Any mercury(II) present would precipitate with sulphide to form mercuric sulphide. However, mercury(II) must be absent when used as a silver electrode. Calibration plots using the Orion 94-16 electrode in pure sulphide, and

---

* This is quite an astonishing limit by any analytical standards. 6.1×10^{-8}g of radium 226, that is, 1.6×10^{14} atoms emit a barely detectable ten α-particles per second.

in eight different mixed anion-sulphide solutions, are virtually identical. It is concluded[22] that "the numerical selectivity ratios cannot be evaluated, and are for practical purposes, infinite" (Table 8). Mercury, silver and sulphide ions are general interferences for most silver based cation, or anion, sensitive solid state electrodes (Tables 7 and 8).

Response Times
Response times depend on stirring, mixing efficiency, and the condition of the membrane surface, but are independent of the initial sulphide, or silver, concentrations.[22,23,94] The $t_{\frac{1}{2}}$ response times following a two-fold increase in the initial sulphide concentration in several sulphide solutions, each between 1 and 6×10^{-5} M, are about one second for the Orion 94-16 electrode. This is shorter than for Pungor type electrodes.[18,20] A steady e.m.f. response takes from 20 to 35 s for the 94-16 Orion electrode depending on initial sulphide concentration,[22] although a separate study[94] indicates a slightly longer time of about 60 s for either a hundred-fold increase, or decrease, in silver concentration. However, when the electrode is first equilibrated with a test sample it takes a mere 5 ms to achieve a constant potential on completion of circuit.[23] The response time of the Beckman silver sulphide electrode varies from a few seconds to several minutes depending on solution viscosity.

Applications of the Orion 94-16 Sulphide Electrode
This electrode can be used like any other electrode in one of two ways:

(i) Direct potentiometric measurements based on a single e.m.f. reading with a pH or specific ion meter (this also covers known addition or subtraction techniques).
(ii) Potentiometric titrations, which give total concentration, and often with greater accuracy than the direct approach since the titration curves can involve such a large e.m.f. change (Fig. 20).

81

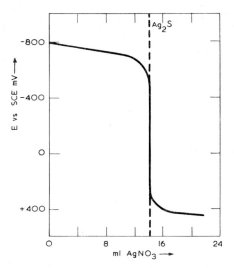

Fig. 20. Typical titration of Na₂S (1·473 × 10⁻²M) with AgNO₃ (1·012 × 10⁻¹M) at μ = 0·1M (Ref. 22).

One technique[95] permits total sulphide levels as low as about 3×10^{-6} M to be determined. Thus, Standard Anti-Oxidant Buffer (S.A.O.B.) comprising sodium salicylate and ascorbic acid, is added (1:1 by volume) to calibrants and samples alike. The ascorbic acid offsets sulphide oxidation, while the buffering action (cf. T.I.S.A.B., p. 71) fixes the pH at a level where the sulphur species is mainly S^{2-}, and not undetectable HS^- or H_2S.

$$HS^- + OH^- \rightleftharpoons S^{2-} + H_2O \qquad (52)$$

The Orion 94-16 electrode has been used[22] to evaluate the solubility product of silver sulphide, $(1·48 \pm 0·1) \times 10^{-51}$, which compares favourably with earlier literature values of $6·2 \times 10^{-52}$ and 7×10^{-50} respectively. Hseu and Rechnitz[22] also determined the formation constant for the equilibrium,

$$SnS_2(s) + S^{2-} \rightleftharpoons SnS_3^{2-} \qquad (53)$$

82

to be $(2 \cdot 062 \pm 0 \cdot 098) \times 10^5$ at 25°, compared with a previous value of $1 \cdot 1 \times 10^5$ at 20°. Sulphide has also been successfully analysed in waters,[95] beer,[96] and industrial solutions, notably the highly alkaline "black liquor" of the pulp and paper trades.[97]

Black liquor comprises mainly inorganic sulphide, some 50 per cent being the ionized S^{2-} species, and the little-ionized organic fraction of mercaptans and methyl sulphides.[97] In practice, spent liquor is purged with oxygen to minimize sulphur losses, and to reduce the release of poisonous sulphur compound vapours. The 94-16 electrode has been used for the continuous on-line monitoring of black liquor processing, where a periodic knowledge of sulphide concentrations is helpful.[97] During the first hour of oxygenation the free sulphide content decreases by some fifteen decades (Fig. 21, Curve A), whereas a platinum electrode merely records about nine decades (Fig. 21, Curve B). However, in the next thirty-minute period the sulphide concentration actually increases by some four decades, whereas the platinum electrode continues to record a further four decade decrease. The sulphide level is finally lowered to about 10^{-18} M after three hours of oxygen bubbling. Despite the lack of resolution achieved with the platinum electrode, the total three-hour decade span is essentially that recorded by the Orion 94-16 electrode (Fig. 21). This is an excellent example of the tremendous future rôle to be played by these new electrodes on the industrial-pollution front.

Heavy Metal Sulphide-Silver Sulphide Electrodes

Three divalent metal(M)electrodes have been marketed whose membranes comprise MS-Ag_2S mixtures. The ultimate silver ion level in any non-silver ion sample, due to the finite membrane solubility, is related to the two equilibria,

$$MS(s) \rightleftharpoons M^{2+} + S^{2-} \tag{54}$$

$$Ag_2S(s) \rightleftharpoons 2Ag^+ + S^{2-} \tag{55}$$

83

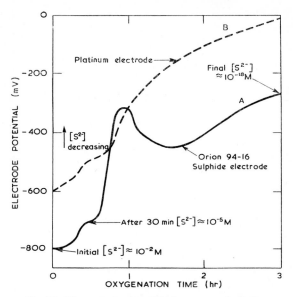

Fig. 21. The variation in sulphide content of alkaline pulp liquor during an oxygen purging process.

Their respective levels in the sample are given by,

$$a_{M^{2+}} = K_{S_{MS}}/a_{S^{2-}} \qquad (56)$$

and equation 50. Combination of equations 50 and 56 gives,

$$a_{Ag^+} = \sqrt{a_{M^{2+}} (K_{S_{Ag_2S}}/K_{S_{MS}})} \qquad (57)$$

The e.m.f., E, of the cell,

| Ag; AgCl(s) | M^{2+}(solution) | MS(s) Ag$_2$S(s) | M^{2+} sample | Reference Electrode |

is, therefore,

$$E = E^\circ + 2 \cdot 303 \frac{RT}{2F} \log a_{M^{2+}} \qquad (58)$$

84

Such a mixed metal sulphide electrode device will record M^{2+} sample activities in a Nernstian manner, and without interference from oxidizing and reducing agents, provided,[98]

(i) the divalent metal sulphide is more soluble than silver sulphide, that is, $K_{s_{MS}} \gg K_{s_{Ag_2S}}$, in order that the equilibrium,

$$Ag_2S(s) + M^{2+} \rightleftharpoons 2Ag^+ + MS(s) \tag{59}$$

will lie well to the left, and so maintain a surface layer of silver sulphide on the membrane;

(ii) the value of $K_{s_{MS}}$ is not too large, otherwise a considerable level of M^{2+} ions will dissolve from the membrane, and make for a low detection limit (cf. the lanthanium fluoride electrode, p. 66);

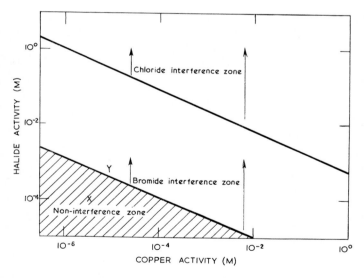

Fig. 22. Halide interference zones for the Orion 94-29 copper electrode.

(iii) the membrane-sample phase boundary equilibrium processes are rapid, so as to solicit favourable response times;

(iv) certain ions, notably mercury(II), are absent in the samples.

The Orion liquid ion exchanger copper and lead electrodes have now been replaced by these superior solid state analogues. Apart from increased mechanical strength, and ability to function in non-aqueous solvents,[24,99] the activity range is extended, by some three decades, for example, in the case of the Orion 94-29 copper electrode. The number of interferences is also drastically cut back to just silver and mercury cations. Any iron (III) ion, which interferes when present at a level of one-tenth, or more, of that of copper, is very easily removed by adjusting the sample to pH > 4 with alkali, but not > 6, otherwise copper hydroxide also precipitates.

Anion interferences can arise in rather a special way,[98] for example, with bromide,

$$Ag_2S(s) + Cu^{2+} + 2Br^- \rightleftharpoons CuS(s) + 2AgBr(s) \quad (60)$$

To maintain the stability of the silver sulphide surface it is essential that

$$a_{Cu^{2+}} \cdot (a_{Br^-})^2 < K_{s_{Ag_2S}}/K_{s_{CuS}} \cdot (K_{s_{AgBr}})^2 < 1 \cdot 3 \times 10^{-12} \quad (61)$$

Thus, there is no bromide interference in a solution comprising Cu^{2+} (10^{-5} M) — Br^- (10^{-4} M), and corresponding to point X (Fig. 22). In a mixture of Cu^{2+} (10^{-5}M) — Br^- (10^{-3}M) however, the product $a_{Cu^{2+}} \cdot (a_{Br^-})^2 = 10^{-11}$ and bromide now interferes as shown by Y (Fig. 22), due to the surface conversion of silver sulphide to silver bromide, and the electrode now functions as a bromide electrode.

Many potential applications have been listed by manufacturers for these mixed sulphide electrodes. The copper model is ideal for measurements in plating and etching baths, rinse tanks, ore refining, industrial effluents, soils, sewage, pharmaceuticals, paints and algaecides.

A few preliminary reports of their performance in non-aqueous solvents and for complexometric titrations have appeared.[24,99-101]

6

NON-GLASS SELECTIVE ION SENSITIVE ELECTRODES: CLASS B. HETEROGENEOUS SOLID STATE ELECTRODES

Pungor has emphasized that the terms homogeneous and heterogenous relate to the composition, and not function, of an electrode.[102] Any theoretical discrimination between the two classes is undesirable, owing to their similar mechanisms.[21] The membrane of a typical solid state electrode (p. 63) is an "all-in one" sensor-cum-support matrix, whereas a heterogeneous membrane comprises at least two components, one the active sensor such as a simple insoluble salt, a chelate or ion exchange resin, and the other an inactive matrix support.

Heterogeneous electrodes include those where the active materials are dispersed in the widely used silicone rubber,[18–21,102–112] as well as polyvinyl chloride[6,17,50,53,113] and paraffin.[50,53,110,112,113] The Shatkay electrodes and two other particular calcium electrodes comprising organophosphate exchanger dispersed in polyvinylchloride and collodion matrices respectively, are really heterogeneous types. Their description and general performances are deferred, however, until Class C on liquid ion exchanger membrane electrodes (p. 94).

A suitable matrix support must:

  (i) be chemically inert and provide good adhesion for the sensor particles;

 (ii) be hydrophobic;

(iii) be tough, flexible, yet non-porous and crack resistant to prevent leakage of internal solutions;

(iv) not swell in sample solutions. This is one reason for the wide use of silicone rubber, which takes up < 0.1 weight per cent of water.[102] Any extensive swelling disrupts the active chain of sensor particles.

In turn, the active sensor material (obtained, for example, by precipitation methods) should:

(i) be physically compatible with the matrix;

(ii) have a low solubility product;

(iii) be of the right grain size $(1 \to 15 \text{ nm})$ which is a function of the precipitation technique;

(iv) be mixed with the matrix support in the right proportion, usually 50 weight per cent. This is essential in order to maintain physical contact between sensor particles, and provide electrical conduction through the membrane, otherwise the resistance may be too great;

(v) undergo rapid ion exchange at the membrane-sample interface.

Preparation of Active Membranes Incorporating Precipitates

Limited details of membrane preparations are available.[19,102,108,110] Silver bromide[19] and lanthanum fluoride[108] materials have been precipitated with excess of cations, and silver chloride[19] and barium sulphate[110] with excess of anions; for silver iodide, excess of precipitant seems unnecessary. The ground, dried precipitate is mixed with silicone rubber monomer and cold polymerized with an added silane to ensure a certain degree of cross linkages in the final membrane $(0.3 \to 0.5 \text{ mm}$ thick). This is related to the particle-matrix adhesion. Buchanan and Seago[110] obtained suitable membranes $(0.5 \text{ mm}$ thick) by pressing a vulcanized silicone-precipitate slurry between a heavy polyethylene plate and polyvinyl chloride sheet.

Fabrication of Heterogeneous Membrane Electrodes

Small discs are cut from these heterogeneous sheet membranes, and secured to one end of a glass tube with a silicone

7

TABLE 20

Pungor-Radelkis Heterogeneous* Ion Selective Electrodes

Addresses
{
Protech, 40 High St., Rickmansworth, Herts (Advisory Services).
Simac Instruments Ltd., Bridgeway House, Bridge Way, Whitton, Middlesex (Technical Services).
Radelkis Electrochemical Instruments, Budapest 62, Hungary.
}

Ion	Laboratory Model†	Molar Range	Maximum Resistance (Megohms)‡	Membrane Material‖	Principal Interferants
Cl^-	OP-Cl-711¶	$10^{-1} \rightarrow 10^{-5}$	10	AgCl	S^{2-} ; I^- ; Br^-
Br^-	OP-Br-711¶	$10^{-1} \rightarrow 10^{-6}$	10	AgBr	S^{2-} ; I^-
I^-	OP-I-711¶	$10^{-1} \rightarrow 10^{-7}$	10	AgI	S^{2-}
S^{2-}	OP-S-711¶	$10^{-1} \rightarrow 10^{-17}$	10	Ag_2S	
SO_4^{2-}	**	$10^{-1} \rightarrow 10^{-5}$		$BaSO_4$	
PO_4^{3-}	**	$10^{-1} \rightarrow 10^{-5}$		$BiPO_4$	
F^-	**			LaF_3-CaF_2	

Operative temperature range (°C) : $1/5 \rightarrow 90$
Overall dimension (length × diameter) cm : 11.5×1 for 711 Models.
Storage : Can be dry stored after a distilled water rinse.
Response time (s) : $15 \rightarrow 60$ (Up to 3 min quoted by Rechnitz and co-workers for some electrodes.)
References 18 and 20.

Cost : $60

* The terms "homogeneous" and "heterogeneous" do not relate to the function, but to the composition of the electrodes. (Pungor, Reference 102).

† Industrial number OP-Ion-722. The OP-Ion-700 series carry side pin electrode connector plugs, while the OP-Ion-711 types are fitted with shielded cabling.

‡ Pure silicone rubber $> 11 \times 10^9$ ohms.

‖ Dispersed in silicone rubber matrix (50 wt per cent of silver salt).

¶ These four electrodes may be used to assay Ag^+ ions after appropriate calibration but do not necessarily cover the same range.

** Not commercially available.

rubber adhesive. The electrode unit is completed with an appropriate internal reference electrode and filling solution. Following a two-hour conditioning by immersion in a 0·1 M alkali metal halide solution, the assembly is ready for use.[21]

Selectivity and Response Times of Pungor-Radelkis Electrodes

Selectivities have been determined in separate[18,20] and mixed solutions.[19,21] (Table 8.) The agreement for certain interferences, notably chloride ion, is excellent for the five iodide and bromide electrodes. Other comparisons are less favourable.

Calibrations of the iodide electrode in potassium, barium and cerium(III) iodide standards are essentially superimposable, indicating the absence of any cation interferences.[18] The bromide and chloride electrodes behave similarly,[20] suggesting that these three membrane electrodes function as anion exchangers. There are slight discrepancies in their linear response ranges,[18,20,102] but in either case the lowest limits lie above those in the range quoted by the manufacturer (Table 20). These limits[18,20,102] are in line with the solubility products of the active membrane constituents.

The e.m.f. is thought to originate through an ion-exchange mechanism at the sample-membrane interface.[20] This is reflected in the increasing $t_{\frac{1}{2}}$ anion response times of the iodide (8 s), bromide (14 s) and chloride (20 s) electrodes, following a two-fold concentration change of the stirred original sample solutions.[20] These responses are independent of concentration between 10^{-2} and 10^{-1} M. Their full-term response times, 2–3 minutes, are considered too sluggish for monitoring purposes in rapidly changing systems.[18]

Applications of Commercial Pungor-Radelkis Electrodes

Pungor has reported them to be useful for determining iodide in mineral waters, cyanide in sewage, brandies and hydrolysed amygdalin samples, chloride in potassium

91

hydroxide (following the removal of hydroxide on a cation exchange resin), and for various complexation and corrosion studies.[102] The chloride content of cheeses can be rapidly found using a Pungor chloride electrode.[111]

Non-Commercial Heterogeneous Electrodes

Macdonald and Tóth have attempted to fabricate fluoride electrodes based on lanthanum, thorium and calcium fluorides embedded in silicone rubber matrices.[108] Particular attention was given to improving precipitation procedures which normally produce unsuitable gelatinous products. Subsequent work with other rare earth fluorides indicates that samarium(III) fluoride membranes are more effective and less subject to interferences.[109]

Three interesting halide electrodes have been constructed from thermoplastic Lucite membranes ($0.3 \rightarrow 1.5$ mm thick) containing $50 \rightarrow 70$ weight per cent of silver halide precipitate.[17] Their linear concentration working range is slightly better than that of the corresponding Pungor-Radelkis types, while the $t_{\frac{1}{2}}$ response time for the Lucite bromide electrode (< 5 s) is shorter than the equivalent time (14 s) of the Pungor-Radelkis electrode.[20] The selectivities of the Lucite iodide electrode compare favourably with those of any commercial iodide electrode (Table 8).

The prototype Pungor-Radelkis sulphate and phosphate electrodes examined in 1967 by Rechnitz, Lin and Zamochnick,[106] are not commercially available items. An approximate Nernstian response is evident using the sulphate electrode in potassium sulphate ($> 10^{-4}$ M), and interference occurs with halides above 10^{-3} M. Even in solutions at pH 12.8, where PO_4^{3-} is the predominant species, calibration plots had low slopes, about 7 mV per concentration decade with the phosphate electrode. Despite these drawbacks they may find limited use for potentiometric titrations.[106]

Since then $AlPO_4$, $CrPO_4$, $FePO_4$, $Co_3(PO_4)_2$, $LaPO_4$, $CoHPO_4$, $[Co(en)_3]PO_4$, $Mg_2P_2O_7$ and Dowex 1-X8 (phosphate form) have each been incorporated into silicone rubber, acrylamide, and paraffin matrices, in an attempt to realize a suitable phosphate electrode.[112] Although the resultant electrodes exhibited a linear e.m.f. response over the concentration range, $10^{-1} \rightarrow 10^{-5}$ M, and gave rapid responses (< 1 min), selectivity could not be achieved.[112] Guilbault and Brignac[112] concluded that "if an ion-sensitive phosphate electrode is to be obtained, a technique other than the precipitate, or ion-exchange, impregnated membranes must be used". Other work[114] with liquid ion exchangers is encouraging (p. 126). Buchanan and Seago have examined silicone rubber electrodes containing anhydrous and hydrated divalent cobalt, copper, nickel and manganese phosphates for their response to the transition metal, but again no selectivity to any one cation is evident.[110]

7

NON-GLASS SELECTIVE ION SENSITIVE ELECTRODES: CLASS C. LIQUID ION EXCHANGER ELECTRODES

There is no shortage of prospective liquid cation and anion exchanger materials for possible use in ion sensitive electrodes. The successful fabrication of such membrane electrodes presents problems, many of which are mechanical. Thus, the liquid ion exchanger:

(i) must be in electrolytic contact with the sample, but any mixing of phases must be minimal;

(ii) must not be too soluble in the sample solution—this factor is related to the lowest detection limits;

(iii) should have a viscosity high enough to prevent its rapid loss by flow across the membrane;

(iv) possess good stability, and be available in a state of high purity with high exchanger capacity. Low cost is another important consideration.

The general specifications of available liquid ion exchanger electrodes are shown in Tables 21–23.

Orion Calcium Liquid Ion Exchanger Electrodes

Several high molecular weight, water immiscible, organo-phosphorus compounds of general formula

$$
\begin{array}{ccc}
\text{Alkyl O} \diagdown \quad \text{O} & \text{O} \quad \diagup \text{O Alkyl} \\
\quad \quad \text{P} & \quad \text{P} \\
\text{Alkyl O} \diagup \quad \text{O—Ca—O} \quad \diagdown \text{O Alkyl}
\end{array}
$$

94

TABLE 21

Beckman Solidion Exchanger Electrodes

Addresses: 2500, Harbor Boulevard, Fullerton, Calif., 92634, U.S.A. (P.O. Box 1, Glenrothes, Fyfe, Scotland)

Ion	Model	Molar Detection Limit	pH Range*	Resistance (Megohms)	Response Time†	Principal Interferants (Where $K_{MN} > 1$)
BF_4^-	39620	10^{-5}	$2 \to 12$	~ 100	Few seconds	ClO_4^-
NO_3^-	39618	10^{-5}	$2 \to 12$	~ 100	Few seconds	I^-; ClO_3^- and ClO_4^-
ClO_4^-	39616	10^{-5}	$2 \to 12$	~ 100	Few seconds	
Ca^{2+}	39608	5×10^{-4}	$3/5 \to 11$	< 500	3 seconds for 90% response	
Divalent $Ca^{2+} - Mg^{2+}$	39614	10^{-5}	$6 \to 11$	~ 100	Few seconds	Zn^{2+}

Electrode life: the easily replaceable sensing element tip is designed to last for at least 100 measuring hours.
Temperature range (°C): $-5 \to 40$, but $-5 \to 50$ for Ca^{2+} electrode.
Overall size, length × diameter (cm) where quoted: 12.8×1.25.
   * No pH effect above 10^{-3} M.
   † But depends on stirring rates, sample concentration and viscosity.

TABLE 22

Some Corning Liquid Ion Selective Electrodes

Addresses: Corning Glass International, Medfield, Mass., 02052, U.S.A. (3 Cork St., London W.1)
(EEL, Hallstead, Essex)

Ion	Model Number	Working Concentration Range	Operating pH Range	Temperature Range (°C)	Principal Interferants ($K_{MN} > 1$)
Cl^-	476131	10^{-1}–10^{-5} M NaCl	1–12 in 10^{-1} M NaCl	10–50	$I^- > ClO_4^- >$ $NO_3^- = Br^-$
NO_3^-	476134	10^0–10^{-6} M KNO_3	2·5–10 in 10^{-2} M KNO_3	0–50	$ClO_4^- > I^-$
Ca^{2+}	476041	10^0–10^{-5} M $CaCl_2$	5–10 in 10^{-3} M $CaCl_2$	10–60	None given
$Ca^{2+} - Mg^{2+}$	476235	10^0–10^{-5} M as Ca^{2+}	5–10 in 10^{-3} M $CaCl_2$	10–60	$(Ba^{2+}=Sr^{2+} >$ $Ni^{2+}=Ca^{2+}=Mg^{2+})$*

Overall size (length × diameter) cm : 12·3/12·7 × 1·58.

Lifetime : 10–15 days (where quoted) before recharging with specified liquid ion exchanger.

Time response (s) : Generally < 60 for solutions differing by not more than ten-fold concentration changes.

Resistance (Megohms) : 500 quoted nominally for Ca^{2+} and $Ca^{2+} - Mg^{2+}$ electrodes only.

Minimum sample size (ml) : 10 quoted for Ca^{2+} and $Ca^{2+} - Mg^{2+}$ electrodes only.

Storage : (Upright in air for short periods.
(Ion exchanger liquid removed for prolonged periods.

* Merely listed as such without any selectivity values.

TABLE 23

Orion 92-Series Liquid Ion Exchange Membrane Electrodes

Addresses: 11 Blackstone Street, Cambridge, Mass., 02139, U.S.A. (E.I.L., Richmond, Surrey)

Ion	Model	Molar Activity Range	pH Range	Resistance (Megohms)	Principal Interferants (where $K_{MN} > 1$)	Mobile Exchanger Site†
Cl^-	92–17	10^{-1}–10^{-5}	2–11	<30	Br^-, I^-, NO_3^-, ClO_4^- and OH^-	NR_4^+
NO_3^-	92–07	10^{-1}–10^{-5}	2–12	<30	I^-, ClO_3^- and ClO_4^-	$[Ni(phen)_3]^{2+}$
BF_4^-	92–05	10^{-1}–10^{-5}	2–12	<30	I^-	
ClO_4^-	92–81	10^0–10^{-5}	4–10/11	<30	OH^-	
Cu^{2+}*	92–29	10^{-1}–10^{-5}	4–7	<30	Fe^{2+}	$R-S-CH_2-CO_2^-$
Pb^{2+}*	92–82	10^{-2}–10^{-5}	3·5–7·5	<10	Cu^{2+}	
Ca^{2+}	92–20	10^0–10^{-5}	5·5–11	<25	Zn^{2+} (But see Table 2)	
Divalent $Ca^{2+}-Mg^{2+}$	92–32	10^0–10^{-8}	5·5–11	<10	Zn^{2+}, Fe^{2+}, Ni^{2+} and Cu^{2+}	$(AlkylO)_2PO_2^-$

Operative temperature range (°C): 0–50.

Overall size, length × diameter (cm): 14·9 × 1·7 (slightly larger dimensions were once quoted).

Minimum sample size (ml): 3–5 in 50 ml beaker: 0·3 in disposable Orion microsample dish.

Reproducibility: Drift, repeatability and response time characteristics are generally comparable with those of good quality pH electrodes.

Storage: Can be air-stored or immersed in appropriate standardized ion solution.

Electrode life: About 1–3 months without renewal of ion exchange liquid.

Dollar cost: 195

* Now withdrawn from Orion 1969 Research Guide. (Cat/961).

† Iron replaces nickel in the ion exchanger material for the ClO_4^- electrode.

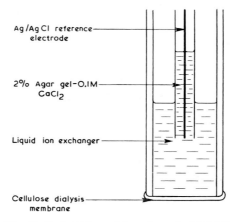

Ag/AgCl reference electrode

2% Agar gel-0.1M CaCl$_2$

Liquid ion exchanger

Cellulose dialysis membrane

Fig. 23. Early Orion calcium electrode (Ref. 98).

have been used[115-118] as liquid ion exchanger materials. These calcium salts are dissolved in a water immiscible mediator such as decanol or di-*n*-octylphenylphosphonate,

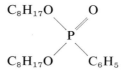

The low dielectric mediator is not merely a solvent but can be employed to adjust:[117]

(i) the ultimate dielectric constant of the final organic phase;

(ii) the mobility of the exchanger sites (R) according to the viscosity of the mediator. The mobilities of the inorganic ions and ion sites are higher than is possible with solid state electrodes;

(iii) the site density by variation of the concentration of the liquid ion exchanger.

A water-immiscible solution of such a system in particular, namely, calcium didecylphosphate (0·1 M) in di-*n*-octylphosphonate, was first used by Ross.[115] One means of maintaining a membrane of this liquid mixture between two aqueous calcium phases is shown in Fig. 23. However, the high resistance and relatively long time response due to the thick layer of exchanger material above the dialysis membrane constituted a drawback, yet the Nernstian slope, time response and selectivities are virtually identical with the later commercial model.

The first commercial calcium electrode to be developed from this prototype model is a complicated piece of fluorocarbon engineering (Fig. 24). It consists of two vertical reservoir

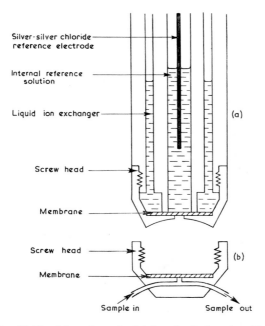

Fig. 24. Orion 92-20 calcium electrode. (a) Longitudinal section. (b) Projection to illustrate flow-through modification.

chambers, a single central one for the internal calcium chloride reference solution (0·1 M), and a double outer pair for the liquid ion exchanger.[118]

The water immiscible ion exchanger-mediator solution is held in a thin millipore filter membrane (0·076 mm thick × 3·5 mm diameter) with "wick" pores (about 100 nm diameter), and which are preferentially "wetted", and filled, by the liquid ion exchanger.[118]

The Orion 92-20 calcium electrode is really a three-phase system wherein the typical aqueous *inner* calcium chloride phase (10^{-3} M) is separated from the *outer* aqueous calcium sample phase by the thin, rigid, immiscible, organic exchanger phase held in the inert porous membrane. Phase mixing is thus minimized. The electrode develops a potential across this thin membrane. The calcium ion in this inner solution maintains a stable potential between the silver–silver chloride internal reference electrode and the inside membrane surface; the chloride ion provides a stable potential between the internal reference electrode and the inner solution. Changes in potential are thus due only to changes in the total sample calcium levels. The e.m.f., E, of the cell,

Ag; AgCl(s)	Ca^{2+}(10^{-3}M)	[(AlkylO)$_2$POO]$_2$Ca (AlkylO)$_2$POC$_6$H$_5$	Test Sample	Reference Electrode

is given simply by,

$$E = E^\circ + 2\text{·}303 \frac{RT}{2F} \log a_{Ca^{2+}} \qquad (62)$$

where E° is the algebraic sum of all potentials except that at the membrane/test sample interface.

The Orion 92-20 calcium electrode exhibits a Nernstian response down to about 5×10^{-5} M. As in the case of solid state and heterogeneous electrodes, the lower detection limit is related to the finite solubility of the calcium exchanger salt in the aqueous sample. This limit (~0·8 p.p.m.) is suitable for most analytical purposes but could be extended by:

(i) increasing the length of the aliphatic chain (alkyl), but precipitation, or gelling, in the organic phase may occur if made too large;

(ii) lowering the amount of calcium salt in the mediator, but this advantage is offset by an increased resistance and lengthened time response.

Deviations from the Nernstian response also appear above about 10^{-1} M calcium chloride.[115] This is due to a large concentration gradient between test and internal calcium solutions. However, by closer concentration matchings of the internal reference solutions and the test sample, it is possible to extend the Nernstian range to even 5 and 6 M calcium chloride solutions.[119]

Generally, only calcium ions can exist in the organic phase, and they are the principal species responsible for carrying charge between sample and ion exchanger. An infinitesimally small number of ions is involved in this transportation process. A high calcium selectivity in the presence of a divalent interferant, N^{2+}, requires that the equilibrium,

$$[(AlkylO)_2POO]_2N + Ca^{2+} \rightleftharpoons [(AlkylO)_2POO]_2Ca + N^{2+}$$
(63)

established at the interface of the membrane-aqueous phase, lies well to the right. Thus, in principle, the liquid ion exchanger should form a stronger complex with calcium than with any other cation.

Selectivity to other alkaline earth metals is reasonable, but not for Zn^{2+}, Pb^{2+} or Fe^{2+} cations (Tables 2 and 24), and where equilibrium 63 will shift to the left. An independent study of the Orion 92-20 electrode[6] shows the selectivity to zinc, and for that matter to hydrogen, to be actually better than the manufacturer's listings (Table 24). Iodide and perchlorate also present[120] interferences above 10^{-3} M owing to their solubility in the ion exchanger. The mediator solvent is said to have little effect on selectivity.

TABLE 24

**Selectivity Constants for Various Calcium Electrodes
Measured in Mixed $Ca^{2+} - N^{n+}$ Solutions**

Interfering Cation N^{n+}	Selectivity Constant for Electrode, K_{CaN}				
	Orion 92-20*†	Orion 92-20‡	PVC‖	Beckman¶ No. 39608	Beckman** No. 39608
Mg^{2+}	0·014	0·04	0·222–0·024	0·12	0·34
Sr^{2+}	0·017	0·07		0·093	
Ba^{2+}	0·01	0·04	0·013–0·004	0·079	0·90
Cu^{2+}	0·27				
Ni^{2+}	0·08				
Pb^{2+}	0·63				
Fe^{2+}	0·80				
Zn^{2+}	3·2		0·065–0·045		
Cd^{2+}		0·03			
K^+	10^{-4}	0·01	$\begin{cases} 3 \times 10^{-5} \\ 2{\cdot}2 \times 10^{-5} \end{cases}$		0·034
Na^+	$\begin{cases} 1{\cdot}6 \times 10^{-3} \\ 3 \times 10^{-3} \end{cases}$	0·01	$\begin{cases} 2{\cdot}1 \times 10^{-4} \\ 5{\cdot}8 \times 10^{-5} \end{cases}$	0·015	0·029
NH_4^+	10^{-4}				
H^+††	$10^5 ; 10^7$		$40 \to 25$	72	$1{\cdot}5 \times 10^4$
$[(C_2H_5)_4N]^+$		10^{-3}		0·46	

* Manufacturer's data. † Ref. 115. ‡ Ref. 9. ‖ Ref. 6. ¶ Ref. 10.
** Ref. 11. †† 2×10^5 for Orme's microelectrode (Ref. 88).
   Calcium selectivity for the Corning electrode (476041) quoted as 100 to 1 over Ba^{2+}; Sr^{2+}; Ni^{2+} and Mg^{2+}; and 1000 to 1 over Na^+ and K^+.

Below about pH 5 the Orion electrode responds to hydrogen, although the slope is not actually Nernstian as depicted in some pH interference plots.[120] The composition of the yellow ion exchanger liquid is not specified, but a mixture of pure calcium didecylphosphate and pure di-*n*-octylphenyl-phosphonate remains water white for months.[121] A characteristic feature of Orion 92-20 electrodes is the dips in their pH interference plots, although these were absent from the original Ross paper.[115] The dips have been attributed[8] to impurities in certain ion exchanger batches, but even material purchased in late 1969 still gave dips.[6] It is interesting that both the Corning and Beckman electrodes, the p.v.c. electrode as well as Orme's microelectrode,[88] all give dips in their pH interference plots. There are no dips in the pH calibration plots for the Orion 92-32 divalent electrode, but they show up with the corresponding Corning divalent electrode. No analogous dips have been recorded for any other type of liquid ion exchanger electrode.

The Orion 92-20 electrode is believed[120] to give correct calcium activity readings above about pH 11, although these actual levels are lowered owing to formation of $CaOH^+$ and $Ca(OH)_2$ complexes. If this is the case, then the proper pH range is $5/6 \rightarrow 14$ and not $5/6 \rightarrow 11$.

The several drawbacks associated with the Orion 92-20 electrode apply to the 92 series in general.

(i) air bubbles tend to get trapped between test sample and membrane. A special electrode holder, which maintains the model 92 series of electrodes at a $20°$ angle out of vertical, is designed to overcome this problem;

(ii) no combination electrode analogous to the Orion 96-solid state series is yet available;[57]

(iii) unlike the solid state electrodes they cannot be used in organic solvent media;

(iv) short "life time" ranging from $2 \rightarrow 3$ weeks for the 92-19 potassium electrode up to 1–3 months for the 92-32 divalent electrode.

The 92-20 model certainly requires refilling at least each month.[53] These short "lifetimes" are related to gradual loss of liquid ion exchangers from the double reservoir through the porous cellulose acetate membrane. (Presumably "lifetimes" would be shorter in flowing systems.)

However, all the Orion 92-models are restored by first discarding the millipore membrane, draining both reservoirs, washing all parts thoroughly in methanol and finally drying. A fresh membrane is carefully refitted (this can sometimes be tricky), the screw head replaced, and both internal reference and liquid ion exchanger syringed into the vertical reservoir chambers. Each electrode is supplied with a kit containing sufficient materials for about twenty such restorations. The short lifetime is of course a "fault" of the design.

However, the body of any Orion-92 model electrode can be adapted to test any commercial, or newly synthesized, liquid ion exchanger liquid. Thus, the particular electrode is completely drained (or, of course, a brand new body taken), and each reservoir loaded with appropriate volumes of internal reference solution and liquid ion exchanger.[12]

Applications of the Orion 92-20 Electrode

Like the lanthanum fluoride electrode, this electrode has been employed in a diverse range of applications.[67] Studies have covered sea water,[122] skim milk,[123] beer,[124] animal feedstuffs,[125] serum,[126] gastric juice,[127] cerebrospinal fluid,[128] clays,[129] equilibrium constants for the calcium-E.D.T.A. and calcium-N.T.A. (nitriloacetic acid) reaction,[9] and the kinetics of the reaction,[9]

$$Ca^{2+} + Mg\text{-}E.D.T.A. \rightleftharpoons Ca\text{-}E.D.T.A. + Mg^{2+} \quad (64)$$

Miniaturization and Modification of Calcium Electrodes

The Orion 92-20 calcium electrode is rather bulky, and minimum sample volumes are limited to around 3–5 ml. However, a special microdish[130] permits analyses on 0·3 ml samples. Orme[88] has designed, but not marketed, a calcium microelectrode, one form of which is shown in Fig. 25. Calibration using a standard reference electrode with liquid junction, gave linear responses over $10° \rightarrow 10^{-5}$ M calcium range and with slopes about $23 \rightarrow 25$ mV per decade. Response times to activity changes in pure calcium chloride solutions are about 20–30 minutes.[88]

An Orion 99-20 flow-through model, of slightly larger dimensions (Fig. 24), now marketed for biological and medical measurements, requires only 200 μl samples. The calcium

104

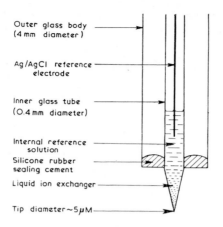

Outer glass body
(4 mm diameter)

Ag/AgCl reference
electrode

Inner glass tube
(0.4 mm diameter)

Internal reference
solution

Silicone rubber
sealing cement

Liquid ion exchanger

Tip diameter ~5 μM

Fig. 25. One type of Orme's (Ref. 88) electrode.

electrode plus the ancillary syringe pump and flow-through reference electrode costs $695. With this complete analytical unit, calcium activity in serum samples can be found within two minutes and under anaerobic conditions.

Other Commercial Calcium Electrodes

Two other electrodes are currently available. All the available data for the three commercial calcium electrodes are listed in Table 25.

The Corning No. 476041 Calcium Electrode
The organic phase of an unspecified nature is held in an inert ceramic plug, which is the equivalent of the Orion millipore membrane (Fig. 26). The electrodes become noisy if the ceramic pores are wetted, and the ceramic pores must be siliconized to prevent wetting by the aqueous phase.[131] A typical calibration plot is shown in Fig. 27.

105

TABLE 25

Specifications of Three Commercial and the PVC Calcium Electrodes

Electrode	Orion 92–20	Corning No. 476041	Beckman No. 39608	PVC (Ref. 6)
Effective concentration range (M)	10^{-5}–10^0	10^{-5}–10^0	5×10^{-4}–10^0	5×10^{-5}–10^{-1}
Effective pH range	5·5–11	5–10	5–11	5–9
Response (mV per decade)	26·5* / 29·58	~30	29·3‖ / 30±1¶	~30
Resistance (Megohms at 25°C)	<25	500 (nominal)	<500	~25
Dynamic response in pure $CaCl_2$ (seconds)	<6† / <30‡	<10	1–10¶ / <10** / 2–3¶ / ~30‖	~6
Effective temperature range (°C)	0–50	10–60	–5–50	0–48
Operational lifetime	30 days	15 days	<100 hours‖ / 90 days‖	>21 months
Size (length × diameter) cm	15 × 1·75	12·7 × 1·6	12·8 × 1·25	11 × 0·7
Approximate cost free of UK import duty (£)	65	40	60	–
Minimum analytical sample required (ml)	3–5††	10	–	0·4

* Ref. 50.   † Refs. 6, 9.   ‡ Ref. 115.   ‖ Ref. 10.   ¶ Ref. 11.

** Time required to obtain a 90 per cent response to a step change of 10^{-4}→10^{-3} M cncentration in a stirred solution.

†† 0·3 ml with special microdish (Ref. 130).

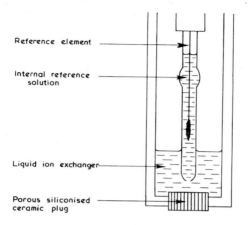

Reference element

Internal reference
solution

Liquid ion exchanger

Porous siliconised
ceramic plug

*Fig. 26. Constructional details of the Corning liquid
membrane electrode.*

The Beckman No. 39608 *Calcium Electrode*

Like the Corning electrode, this electrode is much simpler
in design than the Orion 92-20 model (Fig. 28). No mediator
solvent is used. Either calcium dioctyl, or didecylphosphate
(3–5 parts by weight) is dissolved in an ether-alcohol solution
of collodion. The active organophosphate-collodion matrix
remaining after evaporation of ether-alcohol becomes the
functional calcium sensitive membrane. This electrode
could be classified, like the p.v.c. calcium electrode (p. 113),
as heterogeneous Pungor-type electrodes (p. 88). The Beckman electrode is also of "short" life (Table 25). Restoration
is achieved by taking a complete new solid element unit, and
internal reference solution placed in the narrow glass tubing
sealed above the sensor membrane (Fig. 28). The whole
lower unit is carefully inserted into the electrode body so
that the internal reference wire becomes immersed in the
reference solution, and the assembly secured by tightening
the cap against the electrode body.

107

Little difference might be expected in the performances of these three commercial electrodes. The data collected in Tables 24 and 25 support this idea. There are differences, for example, the Beckman electrode functions properly in 10^{-1} M perchlorate,[10] whereas the Orion electrode cannot tolerate $> 10^{-3}$ M perchlorate.[120] No strict comparison of selectivities is possible until each electrode has been evaluated for one, or more, interferants by identical methods, although again the very limited data show broad interference trends (Table 24).

Particular care is necessary with quoted selectivities owing to the effect of concentration. Thus, the selectivity ratio of calcium over sodium, about 5×10^4 in dilute solutions, is drastically reduced to about 3 in 6 M ionic strength NaCl-CaCl$_2$ solutions.[132] One explanation for this behaviour at

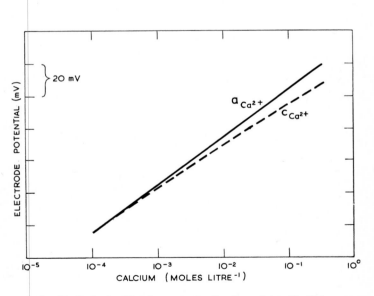

Fig. 27. *Typical calibration curve for Corning calcium liquid ion exchange electrode (as plotted on semilog paper).*

high ionic strengths is based on the participation of the $CaCl^+$ species in the equilibrium

$$CaCl^+ + RNa \rightleftharpoons RCaCl + Na^+ \qquad (65)$$

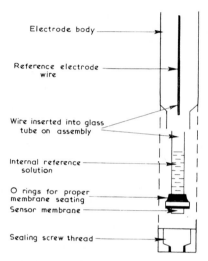

Fig. 28. *Exploded view of Beckman solid ion exchanger electrode.*

since the K_{CaNa} factor depends on the formation of CaR_2 but whose concentration is effectively lowered owing to formation of the $CaCl^+$ complex.[133] This could also account for Nernstian deviation when external and internal calcium levels are widely different.[133]

These liquid ion exchanger electrodes give a better performance when the primary ion concentrations of the internal reference and external sample are of the same order. This fact should be borne in mind when designing, or purchasing, an electrode. Whatever the reasons, only the Orion

92-20 calcium electrode has been widely used, but there seems to be no reason(s) why the other commercial calcium electrodes should not be used with equal success.

Thus the formation constants, K_f, for

$$Ca^{2+} + L^{n-} \rightleftharpoons [CaL]^{2-n} \qquad (66)$$

determined with the Beckman 39068 "solid" membrane electrode are considered to be in "good agreement" with those found by a wide variety of other techniques (second line, Table 26).

TABLE 26

Formation Constants, K_f, for Calcium and Di-; Tri-; and Tetracarboxylic Acid Ligands[10]

Ligand (L)	Malate	Citrate	D.C.T.A.*
Log K_f	2·66 (0·1)†	3·67 (0·1)	12·65 (0·1)
Log K_f	2·01 (0·16)	3·17 (0·15)	13·15 (0·1)

* *Trans*-1,2-diamino-cyclohexane-N,N,N′,N′-tetraacetic acid.
† Parenthesized figures are ionic strengths.

This is one of the few reported studies for a Beckman calcium electrode, which may be used at lower pH levels ($\sim$ 4) than the Orion electrode, athough it is more sensitive to sodium (Table 24). Other advantages listed[10] for the Beckman electrode include a lesser susceptibility to stirring, reduced bubble occlusion and less mechanical disturbance. The accuracy and precision possible with proper use is similar for both electrodes.[10]

Membrane Porosity and Electrode Performance

The multiplicity and the diameter of the membrane pores can influence the stability, time response and selectivity of an electrode.[134] From the limited data of Table 27 it can be tentatively concluded that:

TABLE 27

Some Performance Data for Four Orion Electrodes (One Modified), each containing an Internal Silver-Silver Chloride Reference Electrode and Calcium Didecyl Phosphate and Calcium Didecyl Phosphate in Di-n-octylphenyl phosphonate but with a Different Membrane[134]

Electrode	A	B	C	D*
Membrane material (Organophilic)	Cellulose Acetate	Cellulose Acetate	NaI film D-30[†]	Thick Lexan[‡]
Membrane thickness	130 μm	130 μm	160 μm	~3 mm
Average pore diameter	10 nm	100 nm	6 → 10 nm	1 μm
Percentage hole area in surface	70	70	Very Spongy[‖]	10^5 holes cm^{-2}
σ (mV)[¶]	0·26	0·7	1·1	1·5
K_{CaMg}	0·048	0·068	0·036	0·023
K_{CaNa}	6×10^{-4}	7×10^{-4}	3×10^{-2}	5×10^{-4}
*Transfer response times (min)***				
$Ca^{2+} \leftrightarrow Ca^{2+}$	< 0·1	0·8	1·3	
$Ca^{2+} \to Mg^{2+}$	27	39	3	
$Mg^{2+} \to Ca^{2+}$	0·6	9	5	
$Ca^{2+} \to Na^{+}$	40	40	10	
$Na^{+} \to Ca^{2+}$	40	20	20	

* A simple model made by attaching a Lexan membrane to a pvc tubing and just filled with the above liquid exchanger mixture.

† Believed to be a polyvinyl chloride polymer.

‡ Believed to be a thermoplastic carbonate linked polymer.

‖ Siliconized treatment to render pore linings hydrophobic.

¶ Root Mean Square values of the deviations of the e.m.f. readings in the sample solutions.

** Average times of ten transfers between the sample solutions in the direction(s) arrowed.

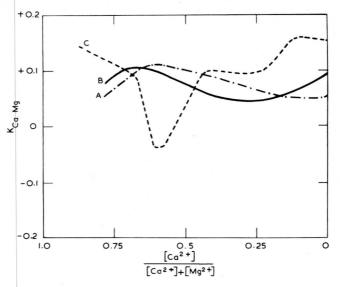

Fig. 29. The dependence of calcium-magnesium selectivity constants on their molar ratios and at constant ionic strength (0·03 M) for three different electrodes (Ref. 134).

(i) electrodes having membranes with a multiplicity of narrow, well interconnected pores have superior stabilities (compare electrodes *A* and *D*, Table 27);

(ii) time responses are related to the pore system of a particular electrode. Each of the four electrodes has been transferred a number of times between various pairs of dilute solutions, and the time noted for the e.m.f. to reach about 98 per cent of its long-term equilibrium value (Table 27);

(iii) selectivity may be dependent on the size of the inter-ferant species. Below a certain pore diameter, it is thought[134] that diffusion of the large CaR_2 species is restricted, whereas the "monovalent" charge mechanism involving smaller molecular species is not impaired.

112

This seems a reasonable explanation for the poor sodium selectivity (by a factor of about $50:1$) of electrode C, compared with about $1400:1$ for either electrode A or B (Table 27). Selectivity is lowered still further (to about $10:1$) for another electrode fitted with a 6 nm pore membrane.[134]

The calcium-magnesium selectivities show far less variation, either between different electrodes (Table 27), or indeed with varying magnesium levels for any one electrode (Fig. 29). A number of $CaCl_2$-$MgCl_2$ solutions, each of total 0.01 M concentration, were prepared and their potentials, E, measured. Knowing E, E°, $a_{Ca^{2+}}$ and $a_{Mg^{2+}}$, the corresponding K_{CaMg} values could be easily found from eqn(10) (p. 12), for each Ca^{2+}-Mg^{2+} system of constant ionic strength ($\mu = 0.03$ M). These selectivities are then plotted against the molar fraction F of each solution,

$$F = [Ca^{2+}]/([Mg^{2+}] + [Ca^{2+}]) \qquad (67)$$

as shown in Fig. 29.

Non-Commercial Calcium Electrodes

Shatkay Electrodes
Electrodes have been constructed using tributyl phosphate and polyvinyl chloride ($3:1$ weight ratio) membranes.[113] Selectivities are quoted for only two interferants, and as (reciprocals of K_{CaN}) $K_{MgCa} = 8.7$ and $K_{NaCa} = 2.5$. Sodium interference is considerably reduced, $K_{NaCa} = 180$, when this membrane matrix is modified to include a third compound, thenoyltrifluoroacetone.[113] Their 30-min response times are a drawback.[53]

The PVC Electrode
Attempts to make a viable calcium electrode from membranes of pure paraffin wax, paraffin wax-nylon gauze, paraffin wax-Zeo Karb 225 or permaflex C20 met with little success.[53] Another choice, incorporating tributylphosphate and didecylphosphoric acid in polyvinyl chloride ($1:3:1$), had a slope of 28.6 mV per decade over a linear activity range 10^{-2} to

10^{-5} M, but an unrealistic 75-min response time.[53] However, a calcium electrode (hereafter abbreviated to pvc electrode) superior in life-time performance to any present commercial item, is easily made[6] from a membrane (0·5 mm thick) of polyvinylchloride, calcium didecylphosphate and di-n-octylphenylphosphonate (Fig. 30). Membranes do not function without the mediator.[121]

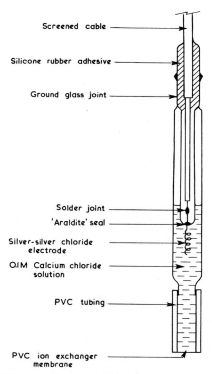

Screened cable

Silicone rubber adhesive

Ground glass joint

Solder joint

'Araldite' seal

Silver-silver chloride electrode

0.1M Calcium chloride solution

PVC tubing

PVC ion exchanger membrane

Fig. 30. Constructional details of the PVC electrode (to scale) (Ref. 6).

The relative performances of the pvc and Orion 92-20 electrodes are given in Tables 2, 24, 25 and 28 and Fig. 31. Both electrodes give steady responses in pure

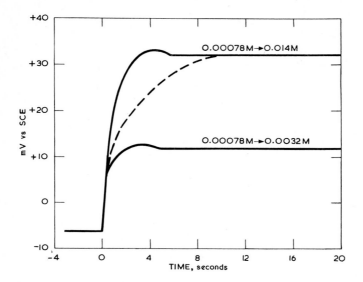

Fig. 31. Dynamic responses of the PVC and Orion 92-20 electrodes in pure calcium solutions. (Activity values shown.)

calcium chloride solutions within about six seconds (full lines in Fig. 31). In the presence of $2 \cdot 5 \times 10^{-2}$ M magnesium chloride, the pvc response time is about 12 s (depicted by broken lines in Fig. 31) whereas the Orion 92–20 electrode is still about 6s. Rechnitz and Lin[9] have reported longer response times for the Orion 92-20 electrode in mixed calcium-magnesium chloride solutions.

The selectivities evaluated under similar conditions by four methods for each electrode show fair pair correlation.[6] In fact, the Orion electrode selectivity for zinc and hydrogen is better than manufacturer's listings (Table 2).

Unlike any other interference examined, zinc has a peculiar effect. Both electrodes take around 1 h to recover their

115

TABLE 28
Known Addition Analysis of Various Calcium Systems[53]

Ion	Solution Volume (ml)	Concentration (M)	Calcium Added (10^{-1}M) (ml)	Orion 92–20 ΔE(mV)	Orion 92–20 $[Ca^{2+}]^*_{Total}$	PVC ΔE(mV)	PVC $[Ca^{2+}]^*_{Total}$	Expected $[Ca^{2+}]_{Total}$
Ca^{2+}	25	10^{-3}	0·05	2·2	$1·056 \times 10^{-3}$	2·3	$1·008 \times 10^{-3}$	1×10^{-3}
			0·10	4·0	$1·077 \times 10^{-3}$	4·0	$1·077 \times 10^{-3}$	
			0·15	5·6	$1·081 \times 10^{-3}$	5·7	$1·056 \times 10^{-3}$	
Ca^{2+}	10	10^{-3}	0·05	5·1	$5·1 \times 10^{-4}$	5·1	$5·1 \times 10^{-4}$	5×10^{-4}†
			0·10	8·6	$5·17 \times 10^{-4}$	8·6	$5·17 \times 10^{-4}$	
			0·15	11·4	$5·21 \times 10^{-4}$	11·3	$5·22 \times 10^{-4}$	
Mg^{2+}	10	10^{-3}						
Ca^{2+}	10	10^{-3}	0·05	3·6	$7·66 \times 10^{-4}$	3·7	$7·41 \times 10^{-4}$	5×10^{-4}‡
			0·10	6·6	$7·34 \times 10^{-4}$	6·4	$7·64 \times 10^{-4}$	
			0·15	8·9	$7·40 \times 10^{-4}$	9·0	$7·28 \times 10^{-4}$	
Mg^{2+}	10	10^{-2}						
Ca^{2+}	10	10^{-3}	0·05	5·0	$5·21 \times 10^{-4}$	5·1	$5·09 \times 10^{-4}$	5×10^{-4}
			0·10	8·7	$5·11 \times 10^{-4}$	8·5	$5·28 \times 10^{-4}$	
			0·15	11·3	$5·25 \times 10^{-4}$	11·2	$5·32 \times 10^{-4}$	
Na^+	10	10^{-2}						

	n	Added Ca²⁺ (ml)	Ca²⁺ 10 · Na⁺ 10	10⁻³ · 10⁰			
		0·05	2·2	1·32 × 10⁻³	2·1	1·38 × 10⁻³	5 × 10⁻⁴
		0·10	3·8	1·42 × 10⁻³	3·8	1·42 × 10⁻³	
		0·15	5·2	1·47 × 10⁻³	5·2	1·47 × 10⁻³	
Cardiff Tap Water	25	0·05	3·3	6·79 × 10⁻⁴	3·2	7·0 × 10⁻⁴	7·5–8 × 10⁻⁴ ‖
		0·10	5·8	6·94 × 10⁻⁴	5·7	7·08 × 10⁻⁴	
		0·15	8·0	6·86 × 10⁻⁴	8·0	6·86 × 10⁻⁴	
Human	15	0·1	3·1	2·4 × 10⁻³	—	—	
		0·2	5·2	2·6 × 10⁻³	—	—	
Serum	10	0·1	—	—	4·3	2·4 × 10⁻³	¶
		0·2	—	—	6·9	2·7 × 10⁻³	

* Using the more exact equation 68 which allows for the volume of added calcium solution.
† No magnesium interference. (See point X on Figure 7.)
‡ Magnesium interference. (See point Y on Figure 7.)
‖ See p. 118 for direct method results, but note that these are taken at a different time.
¶ Frequently quoted for normal subjects at around 2·5 × 10⁻³ M. (Flame photometer value: 2·9 × 10⁻³ M.)

proper potential responses to pure calcium chloride solutions after use in either pure zinc or mixed zinc-calcium chloride solutions, compared with around 10 min for the other interferences studied.[6] Moreover, the e.m.f.s for zinc-calcium chloride solutions are more positive than in pure calcium chloride. The Orion 92-19 potassium[93] and the Orion 92-81 perchlorate[5] electrodes behave similarly in certain mixed solutions.

Some preliminary comparative studies have been undertaken[6] with both the Orion 92-20 and pvc electrodes (Table 28).

The mean calcium (ppm) content of Cardiff tap water determined over a 2-week period by simple EDTA titration (31·62), compared very favourably[6] with the direct electrode values, Orion 92-20 (31·51) and pvc (31·55).

Several mixed calcium-interferant solutions of known concentration have been analysed[6] so as to compare the validity of the basic assumptions listed for the known addition method (p. 54). Provided the magnesium interference is such that the response is still Nernstian (point X, Fig. 7), the expected agreement is good ($\sim$ 1 per cent error). However, the total calcium content is much higher than expected when the magnesium interference arises in the plateau region (point Y, Fig. 7).

The total calcium concentrations in Table 28 are not calculated from eqn(35), but with a somewhat more complicated version,

$$C_o = \frac{C_s}{10^{\Delta E/S}\left(1 + \dfrac{V_o}{V_A}\right) - \dfrac{V_o}{V_A}} \tag{68}$$

Unlike eqn(35), this new equation allows for the volume change following known addition. This results in lower, i.e. more accurate, values for total calcium (Table 28) than would be the case using the simpler eqn(35). The double and triple increment techniques[56] give comparable results.

118

The few total calciums for normal human serum, obtained with either the p.v.c. or Orion 92-20 electrodes,[53] lie approximately within the range of another study[126] for twenty-seven subjects, namely, $2 \cdot 2$–$2 \cdot 67 \times 10^{-3}$ M.

Serum samples of less than 0·4 ml can be directly analysed in a small test tube with a p.v.c. electrode and micro reference electrode.[53] The ionic calcium level of about $1 \cdot 2 \times 10^{-3}$ M, so obtained, compares favourably with the $1 \cdot 3 \times 10^{-3}$ M figure using the Orion microdish arrangement. Again, both these ionic calcium levels lie within the range of $1 \cdot 02$–$1 \cdot 62 \times 10^{-3}$ M quoted for the same twenty-seven patients.[126]

The Divalent (Water Hardness) Electrode

The Orion 92-32 divalent cation electrode was primarily intended for measuring water hardness (defined as the total concentration of divalent cations). It is considered particularly useful in various aspects of water hardness control, for example, corrosion and scale formation, in water softening plants and oceanographic research. Cummins has determined the sesquestering capacities of aminoalkylidene-phosphonic acids by potentiometric titration with standard calcium and magnesium solutions using this divalent electrode.[136]

There is no magnesium selective electrode. Mowbray has mentioned the use of the Corning divalent electrode to obtain total calcium and magnesium, and hence the determination of magnesium by subtracting the calcium level found with the Corning calcium electrode.[137] Thompson has also used the Orion electrode for determining magnesium.[138]

The particular liquid ion exchanger material is uncertain, but di-2-ethylhexylphosphoric acid does not appear to discriminate between magnesium and calcium cations.[117] Didecylphosphoric acid in decanol, which shows virtually identical response to the alkaline-earth metal ions, and yet good selectivity towards monovalent cations, is another possible liquid ion exchanger candidate. The fact that didecylphosphoric acid in di-*n*-octylphenylphosphonate does

119

show good calcium selectivity illustrates that a mediator can sometimes influence selectivity.

In "pure" calcium-magnesium solutions the emf, E, is given by:

$$E = E° + 2.303 \frac{RT}{2F} \log [a_{Ca^{2+}} + a_{Mg^{2+}}] \qquad (69)$$

Except for a much lower detection limit (10^{-8} M), with the Orion 92-32 electrode, the performances of the three commercial divalent electrodes seem fairly similar. Unlike the Corning divalent electrode, and the other calcium electrodes already mentioned, the Orion divalent electrode has no characteristic pH dips. Whatever the cause of these dips, they do shorten the effective pH range of the electrodes by at least a pH unit.

The actual pH range depends on which divalent cation(s) are present since complexation with hydroxyl ions is variously pH dependent. Thus, $MgOH^+$ and $CaOH^+$ formation in their pure 10^{-3} M chloride solutions commences around pH 10 and 11 respectively.

For a calcium-magnesium mixture, and an interferant, say sodium, the e.m.f. is given by,

$$E = E° + 2.303 \frac{RT}{2F} \log [a_{Ca^{2+}} + a_{Mg^{2+}} + K_{CaMg\cdot Na} (a_{Na^+})^2]$$

$$(70)$$

A sodium level 2·6 times that of a 10^{-2} M calcium-magnesium mixture only produces a one per cent error in the electrode response.

There are no electrodes for barium, strontium, nickel or zinc, but any 92-Orion series body can be adopted for their assay, provided other divalent cations are absent, and that less than ten times as much monovalent cations are present.[93] Sufficient divalent liquid ion exchanger, diluted with chloroform, is converted to the appropriate form of the cation of interest, say zinc, by treatment with zinc chloride. The "new" zinc ion exchanger liquid, and internal zinc chloride reference solution (10^{-3} M), are then syringed into the respective reservoirs of a clean 92-series body carrying a fresh millipore membrane.[93]

Orme has also made a few microelectrodes based on the Orion divalent ion exchanger liquid whose slopes, and detection ranges, are almost identical with the calcium microelectrode.[88] Their sensitivity to hydrogen ion is better than the calcium microelectrodes, but about ten times worse for sodium.[88]

Potassium and Other Liquid Ion Exchanger Cation Electrodes

Certain macrocyclic antibiotics exhibit a preference for potassium ions, that is, they have larger formation constants for potassium than, say, sodium. Stefanac and Simon[139] first realized their possible adoption for potassium electrodes. Saturated solutions of nonactin, monactin, dinactin or trinactin in carbon tetrachloride or benzene were examined using glass sinter membranes. Other membrane materials such as filter paper, polyethylene film, Thixcin and nylon mesh were unsatisfactory.[139] Despite low calibration slopes (~ 30 mV per decade), potassium to sodium selectivities up to about 750 were obtained, compared with about 20 for potassium glass electrodes (Table 10). Membranes of low porosity, and antibiotic solutions of high viscosities, gave the best selectivities.[139]

Subsequently, several reports on potassium selective electrodes have appeared.[93,140–148] Other potassium selective materials studied include monensin,[145] nigericin,[145] valinomycin[140,145] and cyclopolyethers (crown compounds).[145] Valinomycin, is used in the Orion and Philips electrodes because it forms a more stable potassium complex than any other antibiotic. Two other potassium electrodes are marketed (Table 29), but no information is available on their particular exchangers.

The Orion 92-19 electrode incorporates a 92-series body with five to ten weight per cent solution of valinomycin in chloro, bromo, or nitrobenzene solvent, and an internal potassium reference solution. A double reference electrode containing lithiumtrifluoroacetate (5 M) in the outer sleeve is employed to prevent potassium contamination of test samples.[144] One drawback of both the Orion and Philips electrodes is their short "lifetimes". The Beckman solid state heterogeneous electrode, valinomycin in collodion (?), should require less frequent recharging; as in the case of the Corning electrode, no such information is quoted in data sheets.

There seems to be no major difference in the overall specifications of the present potassium electrodes (Table 29). Caesium and rubidium constitute the more serious interferences, but are not normally encountered. Lal and Christian[143] have shown interferences for several cations to be very concentration dependent.

The hydrogen selectivity of the Orion electrode is quite poor[143] compared with the other three potassium electrodes, and it is claimed[142] that the Beckman electrode could function without difficulty in unbuffered solutions.

The Philips electrode shows no response at all to chloride, bromide, iodide, cyanide, nitrate or sulphate,[140] but iodide, hydroxide, oxalate and chromate have a slight effect on the Orion electrode.[143]

The time responses quoted need careful interpretation, since an independent study shows them to increase from about one minute at the 10^{-1} M potassium level to between 10 and 15 min for 10^{-4}–10^{-6} M potassium solutions.[143]

Few applications have been reported. The blood serum analysis for eleven patients ranged from 4–5·7 × 10^{-3} M

122

TABLE 29

Specifications* of Four Commercial Potassium Ion Selective Electrodes

Electrode	Beckman 39622 Ref. 142	Corning 476132	Philips 560-K Ref. 140	Orion 92-19¶			
				Ref. 93	Ref. 146	Ref. 144	Ref. 143
Linear molarity range	10^{-1}–10^{-5}	10^0–10^{-4}	10^{-1}–10^{-5}	10^0–10^{-6}	10^0–10^{-6}	10^0–10^{-6}	10^{-1}–10^{-4}
pH Range	2–12†	1·3–11†			1–12**		5–8‡
Temperature range (°C)		15–50	0–50	0–50	0–50		
Resistance (Megohm)	100–500		2	30	250		
Life (weeks)			2	2–3	2–3		
Length × Diameter (cm)		12·3 × 1·58			14·9 × 1·7		
Response time (s)	4–6	60	60			60	60‖ ‖‖
Cost		£60	£60	{$250 / {$480(98-19)			
Selectivity K_{KM} *to:*							
Cs⁺	0·5‡	20‖	0·38	1·0	1·0	1·0‖	0·5¶‖(4)
Rb⁺	2·2‡	10	1·9				2·9(25)
Na⁺	5×10^{-5}	0·012	$2\cdot6 \times 10^{-4}$	2×10^{-4}	2×10^{-4}	7×10^{-4}	0·09(0·02)
Li⁺	3×10^{-4}‡	4×10^{-3}	$2\cdot1 \times 10^{-4}$	10^{-4}	10^{-4}	††	3×10^{-2}
NH₄⁺	0·019	0·023	0·012	0·03	0·03	0·02	0·05(19)
H⁺			$5\cdot5 \times 10^{-5}$	0·01	0·01	0·01	4×10^{-2}
Tl⁺	2×10^{-4}	(See †)			0·01		0·09(0·7)
Ag⁺	$1\cdot7 \times 10^{-3}$‡		2×10^{-9}	10^{-3}	10^{-3}		2×10^{-3}(0·5)
Mg²⁺	2×10^{-5}		2×10^{-4}				10^{-3}
Ca²⁺	2×10^{-5}	3×10^{-3}	2×10^{-5}			2×10^{-4}	$2\cdot6 \times 10^{-3}$
Ba²⁺		5×10^{-3}	6×10^{-5}			2×10^{-4}	
Fe²⁺			4×10^{-4}				
Cu²⁺	3×10^{-5}						$1\cdot3 \times 10^{-3}$

* Manufacturers' data unless otherwise quoted.
† In 10^{-1} M KCl pH range 1·3–11.
‡ Separate solution method.
‖ All figures given in this column are reciprocals of the original values determined using equation 8.
¶ Minimum sample size 0·3 ml in microdish.

** In 10^{-3} M KCl.
†† No response at all to lithium.
‡‡ In 10^{-2} M K⁺.
‖‖ In 10^{-2} M K⁺.
¶¶ Measured in 10^{-3} M interferant solution. Parenthesized values for 10^{-2} M interferant solution.

using either a flame photometer or an Orion 92-19 electrode.[144] This shows that potassium, unlike calcium, is present essentially in ionic form in blood serum. These non-glass potassium electrodes are likely to make the same clinical impact as the calcium electrode. Considerable analytical benefits should follow in other fields, for example, the food and petroleum industries, as well as in oceanography.[93] A flow through model handling 100 μl samples is also available.[93]

The Orion 92-29 copper and 92-82 lead electrodes have been withdrawn,[57] and replaced by the solid state 94-29 and 94-82 models which are superior in all respects.

Anion Liquid Ion Exchanger Electrodes

The essential requirement for this type of electrode is a stable, high molecular weight, water-immiscible ion exchanger with positively charged mobile sites, R^+. The equilibrium,

$$RN + M^- \rightleftharpoons RM + N^- \qquad (71)$$

should lie well to the right. Most of the Orion anion liquid exchangers are highly substituted alkyl or aryl phenanthroline ligands (Table 23).

The Nitrate Electrode

The three commercial nitrate electrodes have similar specifications. (Tables 5 and 21–23.) Iodide, perchlorate and tetrafluoroborate constitute the most serious interferences. The Orion 92-07 electrode has been successfully used in the analysis of waters, soils and plant extracts, as well as for potentiometric titrations and complexation studies.[67]

The Orion 92-05 Tetrafluoroborate Electrode

The tetrafluoroborate ion closely resembles the nitrate and perchlorate ions in several respects, and Carlson and Paul[12] reasoned that the phenanthroline ion exchanger liquid would make a suitable tetrafluoroborate electrode. For this

124

purpose the 92–07 Orion nitrate electrode is easily modified. The internal nitrate solution is replaced by

$$H_3BO_3 \, (10^{-2} \text{ M}); \, HF \, (0.28 \text{ M}); \, \text{and} \, KCl \, (10^{-2} \text{ M}),$$

which forms tetrafluoroborate *in situ*.

$$H_2BO_3^- + 4HF \rightarrow BF_4^- + 3H_2O \qquad (72)$$

The nitrate ion exchanger liquid (2 ml) is converted to the tetrafluoroborate form by shaking with two separate 50 ml lots of tetrafluoroborate solution.

There is no boron electrode as such, but provided the boron in any analytical sample can be quantitatively converted to tetrafluoroborate, this electrode holds considerable promise for boron assay of water,[12] steel[12] and agricultural samples.[149]

The boron sample is introduced to a column of boron specific Amberlite XE-243 resin. Boron is complexed by the *N*-methylglucamine groups and any interfering nitrate or iodide removed by washing with dilute ammonia.[12] Some similar means of removing any interfering perchlorate in samples would be necessary.

Boron is then rapidly converted to tetrafluoroborate by treating the column with hydrofluoric acid, water to remove excess acid, and finally sodium hydroxide to release the tetrafluoroborate. This eluate is passed through a cation exchanger (hydrogen form) to provide tetrafluoroborate in dilute hydrofluoric acid, a form ideal for electrode assay.[12]

Both the Orion and Beckman tetrafluoroborate electrodes have similar selectivities (Table 5).

The Orion 92-17 Chloride Electrode
Unlike the analogous copper and lead ion exchanger electrodes, this chloride electrode, based on dimethyl-distearyl ammonium cation exchanger, has not become obsolete with the advent of the Orion solid state 94-17 model.[57] This is because the Orion 92-17 and 94-17 chloride models each possess certain selectivity advantages. Thus, the 92-17

125

chloride electrode can tolerate sulphide, and small amounts of bromide and iodide, all of which are serious interferants for the solid state chloride electrode (Table 8).

Orme[88] has made a microchloride electrode (Fig. 25) which functions at high chloride concentrations. Cations present no interferences, and the micro electrode works equally well in HCl, LiCl, NaCl, KCl, CsCl, NH_4Cl or $CaCl_2$. Normal response times are about 20 minutes. After exposure to bisulphide, a serious interferant, a micro chloride electrode can take five to ten hours to recover its potential composure.[88]

Non-Commercial Liquid Anion Exchanger Electrodes

It has not been possible to construct a viable heterogeneous phosphate electrode.[106,112] However, two liquid amine exchangers have been employed[114] as their chloride salts, and without mediator solvents, in 92-20 model Orion bodies (Table 30).

TABLE 30

Some Characteristic Features of Divalent Phosphate Electrodes[114]

Ion Exchanger (Chloride Salt)	Internal Reference Solution*	Linear Activity Range (M) (at pH 7–7·5)	Slope mV $(\log a_{HPO_4^{2-}})^{-1}$
Primary Amine	NaCl (0·025 M)	$\begin{cases} 2 \times 10^{-4}\text{–}10^{-2} \\ 2 \times 10^{-3}\text{–}10^{-2}\dagger \end{cases}$	Nernstian 33
Quaternary Amine	$NaHCO_3$ (0·1 M)	$5 \times 10^{-4}\text{–}10^{-2}\dagger$	30

* Also used for conditioning electrodes.
† In the presence of 10^{-2} M chloride.

The electrodes with chloride salts showed better selectivity and sensitivity for divalent phosphate (HPO_4^{2-}) than the corresponding phosphate salts, and had response times of two to five minutes.[114] In the absence of chloride, both electrodes responded down to about 10^{-5} M divalent phosphate levels.

TABLE 31

Some Specifications of Sixteen Non-Commercial Liquid Ion Exchanger Electrodes [150, 151]

Electrode	Concentration Range		Slope* (mV per $\log a_M$)	Selectivity Constants (K_{MN})†								
	Linear	Useful		Cl^-	NO_3^-	SO_4^{2-}	I^-	Br^-	ClO_4^-	ClO_3^-	OBz^-	p-tosate
ClO_4^-	10^{-1}–10^{-3}	10^{-1}–10^{-4}	59.2	7.9×10^{-3}	~3.16‡	0.02				0.16		
Cl^-		10^{-1}–10^{-5}	56.0	~0.16	2.14	0.04	~0.02					
Br^-			59.0	~0.16	0.85	0.03		~3.16				
I^-		10^{-1}–10^{-4}	59.0	~0.01	0.19	6.3×10^{-3}		~0.1				
NO_3^-			57.0	0.23						0.89		
SCN^-			58.0	$<3.2 \times 10^{-4}$	0.06	2.5×10^{-3}	0.28		0.54			
SO_4^{2-}	10^{-2}–10^{-4}		40.0	39.8	10^3	7.9×10^{-3}						
Oxalate	10^{-1}–10^{-3}	10^{-1}–10^{-5}	40.0	~1.0		~0.31					~31.6	~1.59
Formate			53.0	1.32							~5.01	
Acetate			53.0	1.95	~5.01	0.16					~5.01	~1.2
Propionate			57.5	1.45							~3.16	
Benzoate		10^{-1}–10^{-4}	58.6	~0.5	0.79							
Salicylate		10^{-1}–10^{-3}	56.0	$<10^{-3}$	0.05	6.3×10^{-3}						
m-Toluate		10^{-1}–10^{-4}	58.0	~0.1	0.01						0.48	
p-Toluate		10^{-1}–10^{-5}	57.0	0.04	0.016						0.74	
p-Toluene-Sulphonate		10^{-1}–10^{-4}	57.0	0.03	~0.2				2.51		~0.1	0.13

* mV against log concentration for organic electrodes.

† Originally expressed as log K_{MN} (Ref. 151).

‡ The approximation sign, ~, indicates the selectivity to be markedly dependent on the concentration of the interferant.

Sixteen organic and inorganic salts of methyltricaprylyl ammonium ion (Aliquat 336S) in l-decanol mediator have been examined as prospective liquid ion exchangers using Orion model 92-20 electrode bodies.[151,152] Typically the Aliquat salt of, say, oxalate in l-decanol, and reference solution (0·1 M sodium chloride + 0·1 M sodium oxalate), are syringed into the appropriate electrode reservoirs. After conditioning for two to three hours in sodium oxalate (0·1 M), a constant potential is attained within twenty to sixty seconds. Such an electrode lasts up to two months before needing a recharge.

Except for the two divalent anion electrodes, the potential responses of these electrodes (Table 31) are essentially Nernstian. Many have a useful 10^{-1}–10^{-4} M or 10^{-5} M range, but the linear range is restricted to two decades. The fact that the potential response of the iodide electrode is independent of the Aliquat concentration, at least between one and twenty-five volumes per cent,[151] could have an important bearing on cost reduction of liquid ion exchanger electrodes.

The selectivities given in Table 31 for the perchlorate, chloride and nitrate electrodes compare favourably with their commercial counterparts (Table 5). Some (but not all) of the selectivities are essentially independent of interferant concentration.[151] Good organic-ion selectivities are not evident, but future work along these lines[150,151] holds definite prospects for improved organic-ion sensitive electrodes.

This type of study [114,150,151] illustrates the utility of the model 92 Orion casing for anyone wishing to investigate any ion exchanger materials for new electrodes.

8

MISCELLANEOUS ION ANALYSIS

It is possible to determine ions for which there is no selective electrode, either directly or by potentiometric titration.

(1) An ion-exchanger can be chemically modified as already discussed for the Orion 92-07 nitrate to tetrafluoroborate electrode (p. 124), and the Orion 92-32 divalent cation to strontium, barium, zinc or nickel electrodes (p. 120).

A quite different procedure is the separate use of two specific electrodes for magnesium assay in a magnesium-calcium mixture.[137]

(2) A much wider range of ions can be covered by potentiometric titrations. The essence of the technique is to use an ion selective electrode for an ion which reacts stoichiometrically with the one of analytical interest to give a precipitate.

Thus thorium,[72] aluminium[152] and lithium[153] have all been determined by titration against standard fluoride solutions with a solid state fluoride electrode.

The solid state lead electrode has also been employed to determine sulphate, phosphate, tungstate (WO_4^{2-}) and molybdate (MoO_4^{2-}).[100,154] When the solubility product of the precipitate is too large, the titration end-point may be poor. However, the addition of a water miscible organic solvent to the analytical sample before titrating, for example, acetone, dioxan or methanol (20–70 per cent) effectively lowers the solubility product, thereby sharpening the titration end point break. Unlike the liquid ion exchanger electrodes, only solid state electrodes could be used in such media.

129

REFERENCES

1. G. A. Rechnitz, *Anal. Chem.*, **41**, 109A (1969).
2. F. Helfferich, *Ion Exchange*. McGraw-Hill: New York, 1952.
3. P. Debye and E. Hückel, *Phys. Z.*, **24**, 185 (1923).
4. C. W. Davies, *J. Chem. Soc.*, 2093 (1938).
5. K. Srinivasan and G. A. Rechnitz, *Anal Chem.*, **41**, 1203 (1969).
6. G. J. Moody, R. B. Oke and J. D. R. Thomas, *Analyst*, **95**, 910 (1970).
7. Orion Research Inc., *Newsletter*, **1**, 29 (1969).
8. M. S. Frant quoted by F. W. Orme in *Glass Microelectrodes* (Eds. M. Lavallée, O. F. Schanne and N. C. Hébert). Wiley: New York, 1968, p. 384.
9. G. A. Rechnitz and Z. F. Lin, *Anal. Chem.*, **40**, 696 (1968).
10. G. A. Rechnitz and T. M. Hseu, *Anal. Chem.*, **41**, 111 (1969).
11. F. A. Schultz, A. J. Petersen, C. A. Mask and R. P. Buck, *Science*, **162**, 267 (1968).
12. R. M. Carlson and J. L. Paul, *Anal. Chem.*, **40**, 1292 (1968).
13. S. S. Potterton and W. D. Shultz, *Anal. Letters*, **1**, 11 (1967).
14. R. F. Hirsch and J. D. Portock, *Anal. Letters*, **2**, 295 (1969).
15. R. J. Baczuk and R. J. Dubois, *Anal. Chem.*, **40**, 685 (1968).
16. G. A. Rechnitz and T. M. Hseu, *Anal. Letters*, **1**, 629 (1968).
17. M. Mascini and A. Liberti, *Anal. Chim. Acta*, **47**, 339 (1969).
18. G. A. Rechnitz, M. R. Kresz and S. B. Zamochnick, *Anal. Chem.*, **38**, 973 (1966).
19. J. Havas, E. Papp and E. Pungor, *Acta Chim. Acad. Sci. Hung.*, **58**, 9 (1968).
20. G. A. Rechnitz and M. R. Kresz, *Anal. Chem.*, **38**, 1786 (1966).
21. E. Pungor and K. Tóth, *Anal. Chim. Acta*, **47**, 291 (1969).
22. T. M. Hseu and G. A. Rechnitz, *Anal. Chem.*, **40**, 1054 (1968).
23. T. S. Light and J. L. Swartz, *Anal. Letters*, **1**, 825 (1968).
24. M. J. D. Brand, J. J. Millitello and G. A. Rechnitz, *Anal. Letters*, **2**, 523 (1969).
25. A. I. Vogel, *Quantitative Inorganic Analysis*, 3rd edition. Longmans: London, 1961, p. 1166.
26. M. Cremer, *Z. Biol.*, **47**, 562 (1906).
27. F. Haber and Z. Klemensiewicz, *Z. Physik. Chem.*, **67**, 385 (1909).
28. W. S. Hughes, *J. Amer. Chem. Soc.*, **44**, 2860 (1922).
29. G. Eisenman (Ed.), *Glass Electrodes for Hydrogen and Other Cations*. Edward Arnold: London, 1967.

130

30. K. Schwabe and H. Dahms, *Monatsber Deut. Akad. Wiss. Berlin*, **1**, 279 (1959).
31. B. P. Nicolsky, M. M. Schultz, A. A. Belijustin, and A. A. Lev, in *Glass Electrodes for Hydrogen and Other Cations* (Ed. G. Eisenman). Edward Arnold: London, 1967, p. 174.
32. G. Eisenman, D. C. Ruskin and J. H. Casby, *Science*, **126**, 831 (1957).
33. G. Eisenman, *Z. Biol.*, **2**, 259 (1962).
34. G. A. Rechnitz, *Chem. and Eng. News*, **45**, 146 (1967).
35. G. A. Rechnitz and J. Branner, *Talanta*, **11**, 617 (1964).
36. G. A. Rechnitz and S. B. Zamochnick, *Talanta*, **11**, 1061 (1964).
37. G. L. Gardner and G. H. Nancollas, *Anal. Chem.*, **41**, 202 (1969).
38. J. F. McClure and G. A. Rechnitz, *Anal. Chem.*, **38**, 136 (1966).
39. G. A. Rechnitz, S. B. Zamochnick and S. A. Katz, *Anal. Chem.*, **35**, 1322 (1963).
40. D. Hawthorn and N. J. Ray, *Analyst*, **93**, 158 (1968).
41. H. M. Webber and A. L. Wilson, *Analyst*, **94**, 209 (1969).
42. H. D. Portnoy and E. S. Gurdjan, *Clin. Chim. Acta*, **12**, 249 (1965).
43. S. M. Friedman in *Glass Electrodes for Hydrogen and Other Cations* (Ed. G. Eisenman). Edward Arnold: London, 1967, p. 442.
44. J. H. Halliday and F. W. Wood, *Analyst*, **91**, 802 (1966).
45. M. L. Richardson, *Talanta*, **15**, 485 (1968).
46. G. A. Rechnitz and S. B. Zamochnick, *Talanta*, **11**, 979 (1964).
47. G. A. Rechnitz and G. Kugler, *Z. Anal. Chem.*, **214**, 405 (1965).
48. J. E. McClure and T. B. Reddy, *Anal. Chem.*, **40**, 2064 (1968).
49. J. Montalvo and G. G. Guilbault, *Anal. Chem.*, **41**, 1897 (1969).
50. A. Shatkay, *Anal. Chem.*, **39**, 1056 (1967).
51. L. Meites (Ed.), *Handbook of Analytical Chemistry*. McGraw Hill: New York, 1963, Section I, p. 6.
52. R. A. Robinson and R. H. Stokes, *Electrolytic Solutions*, 2nd edition. Academic Press. New York, 1963, p. 230.
53. R. B. Oke, Ph.D. Thesis, University of Wales, 1971.
54. Orion Research Inc., *Newsletter*, **1**, 5 (1969).
55. Orion Research Inc., *Newsletter*, **1**, 22 (1969).
56. Orion Research Inc., *Newsletter*, **2**, 5 (1970).
57. Private communication from Orion Research Inc.
58. U.S. Patent No. 3442782, May 6, 1969.
59. U.K. Patent No. 1150698, April 30, 1969.
60. U.S. Patent No. 3492216, January 27, 1970.
61. U.S. Patent No. 3431182, March 4, 1969 (U.K. Patent No. 1131574, October 23, 1968).
62. Canadian Patent No. 763082, July 11, 1967.
63. I. M. Kolthoff and H. L. Sanders, *J. Amer. Chem. Soc.*, **59**, 416 (1937).
64. H. J. C. Tendeloo, *Proc. Acad. Science, Amsterdam*, **38**, 434 (1935).
65. J. W. Ross, International Symposium in Analytical Chemistry, Birmingham, July, 1969.
66. M. S. Frant and J. W. Ross, *Science*, **154**, 1553 (1966).
67. Orion Research Inc., Bibliography, April 15, 1970.
68. M. S. Frant and J. W. Ross, *Anal. Chem.*, **40**, 1169 (1968).
69. N. T. Crosby, A. L. Dennis and J. G. Stevens, *Analyst*, **93**, 643 (1968).

131

70. W. P. Ferren and N. A. Shane, *J. Food Science*, **34,** 317 (1969).
71. W. J. Simpson and J. Tuba, *J. Oral Medicine*, **23,** 104 (1968).
72. J. J. Lingane, *Anal. Chem.*, **39,** 881 (1967).
73. J. J. Lingane, *Anal Chem.*, **40,** 935 (1968).
74. E. W. Baumann, *J. Inorg. Nuc. Chem.*, **31,** 3155 (1969).
75. K. Srinivasan and G. A. Rechnitz, *Anal. Chem.*, **40,** 1818, 1955 (1968).
76. J. Tusl, *Clin. Chim. Acta*, **27,** 216 (1970).
77. P. Grøn, H. G. McGann and F. Brudevold, *Arch. Oral Biol.*, **13,** 203 (1968).
78. D. R. Taves, *Nature*, **217,** 1050 (1968).
79. N. A. Shane and D. Miele, *J. Pharm. Science*, **57,** 1260 (1968).
80. L. Singer and W. D. Armstrong, *Anal. Chem.*, **40,** 613 (1968).
81. C. R. Edmond, *Anal. Chem.*, **41,** 1327 (1969).
82. J. C. Van Loon, *Anal. Letters*, **1,** 393 (1968).
83. T. S. Light and R. F. Mannion, *Anal. Chem.*, **41,** 107 (1969).
84. L. A. Elfers and C. E. Decker, *Anal. Chem.*, **40,** 1658 (1968).
85. H. H. Moeken, H. Eschrich and G. Willeborts, *Anal. Chim. Acta*, **45,** 233 (1969).
86. R. H. Babcock and K. A. Johnson, *J. Amer. Water Works Association*, **60,** 953 (1968).
87. R. A. Durst and J. K. Taylor, *Anal. Chem.*, **39,** 1483 (1967).
88. F. W. Orme, *Glass Microelectrodes* (Eds. M. Lavallée, O. F. Schanne and N. C. Hébert). Wiley: New York, 1968, p. 388.
89. Beckman Instruction Manual No. 81734A for Silver Halide Electrodes, August, 1969.
90. Beckman Select Ion Electrodes Bulletin, No. 7145.
91. L. Hansen, M. Buechele, J. Koroschec and W. J. Warwick, *Amer. J. Clin. Pathol.*, **49,** 834 (1968).
92. L. Kopito and H. Schwachman, *Pediatrics*, **43,** 794 (1969).
93. Orion Research Inc., *Newsletter*, **1,** 13 (1969).
94. T. S. Light and J. L. Swartz, paper presented at the Pittsburg Analytical Chemistry and Applied Spectroscopy Conference, Cleveland, March 3–8, 1968.
95. Orion Research Inc., *Applications Bulletin*, No. 12, 1969.
96. J. L. Owades, R. Blick and S. H. Owades, *A.S.B.C. Proc.*, 75 (1967).
97. T. S. Light, *Industrial Water Engineering*, September, 1969.
98. J. W. Ross, *Ion Selective Electrodes* (Ed. R. A. Durst), Special Publications 314. National Bureau of Standards: Washington, D.C., 1969, p. 80.
99. G. A. Rechnitz and N. C. Kenny, *Anal. Letters*, **2,** 395 (1969).
100. Orion Research Inc., *Applications Bulletin*, No. 11, 1969. J. W. Ross and M. S. Frant, *Anal. Chem.*, **41,** 967 (1969).
101. J. W. Ross and M. S. Frant, *Anal. Chem.*, **41,** 1900 (1969).
102. E. Pungor, *Anal. Chem.*, **39,** 28A (1967).
103. E. Pungor, J. Havas and K. Tóth, *Acta Chim. Acad. Sci. Hung.*, **41,** 239 (1964).
104. E. Pungor, K. Tóth and J. Havas, *Acta Chim. Acad. Sci. Hung.*, **48,** 17 (1966).
105. E. Pungor, K. Tóth and J. Havas, *Mikrochim. Acta*, 690 (1966).

106. G. A. Rechnitz, Z. F. Lin and S. B. Zamochnick, *Anal. Letters*, **1**, 29 (1967).
107. E. Pungor, J. Havas and K. Tóth, *Z. Chem.*, **5**, 9 (1965).
108. A. M. G. Macdonald and K. Tóth, *Anal. Chim. Acta*, **41**, 99 (1968).
109. A. M. G. Macdonald, International Symposium on Analytical Chemistry, Birmingham, July, 1969.
110. E. B. Buchanan and J. L. Seago, *Anal. Chem.*, **40**, 517 (1968).
111. V. H. Holsinger, L. P. Posati and M. J. Pallansch, *J. Dairy Science*, **50**, 1189 (1967).
112. G. G. Guilbault and P. J. Brignac, *Anal. Chem.*, **41**, 1136 (1969).
113. R. Bloch, A. Shatkay and H. A. Saroff, *Biophys. J.*, **7**, 865 (1967).
114. I. Nagelberg, L. I. Braddock and G. L. Barbero, *Science*, **166**, 1403 (1969).
115. J. W. Ross, *Science*, **156**, 1378 (1967).
116. Belgian Patent, No. 668409, February 17, 1966.
117. U.S. Patent, No. 3406102, October 15, 1968.
118. U.S. Patent, No. 3467590, September 16, 1969.
119. R. Huston and J. N. Butler, *Anal. Chem.*, **41**, 200 (1969).
120. Orion Calcium Electrode Instruction Manual.
121. G. Griffiths, G. J. Moody, R. B. Oke and J. D. R. Thomas, unpublished work.
122. M. E. Thompson and J. W. Ross, *Science*, **154**, 1643 (1966).
123. P. J. Muldoon and B. J. Liska, *J. Dairy Science*, **52**, 460 (1969).
124. *Orion Research Inc., Applications Bulletin*, No. 8, 1970.
125. R. D. Allen, J. Hobley and R. Carriers, *J. Association Offic. Agric. Chemists*, **51**, 1177 (1968).
126. D. E. Arnold, M. J. Stansell and H. H. Malvin, *Amer. J. Clin. Path.*, **49**, 627 (1968).
127. E. W. Moore and G. M. Makhlouf, *Gastroenterology*, **55**, 465 (1968).
128. E. W. Moore and A. L. Blum, *J. Clin. Invest.*, **47**, 70A (1968).
129. W. Fertl and F. W. Jessen, *Clays and Clay Minerals*, 17 (1969).
130. U.S. Patent, No. 3467591, September 16, 1969.
131. F. W. Orme, *Glass Microelectrodes* (Eds. M. Lavallée, O. F. Schanne and N. C. Hébert). Wiley: New York, 1968, p. 384
132. E. W. Moore, *Ann. New York Acad. Sci.*, **148**, 93 (1968).
133. M. S. Frant and J. W. Ross quoted in Reference 119.
134. U.S. Patent No. 3438886, April 15, 1969.
135. C. Sachs, A. M. Bourdeau and S. Balsan, *Ann. Biol. Clin.*, **27**, 487 (1969).
136. R. W. Cummins, *Detergent Age*, 22 (1968).
137. J. H. Mowbray, *Laboratory Equipment Digest*, 45 (1970).
138. M. E. Thompson, *Science*, **153**, 867 (1966).
139. Z. Stefanac and W. Simon, *Microchem. J.*, **12**, 125 (1967).
140. L. A. R. Pioda, V. Stankova and W. Simon, *Anal. Letters*, **2**, 665 (1969).
141. Swiss Patent, No. 479870, November 28, 1969.
142. I. H. Krull, C. A. Mask and R. E. Cosgrove, *Anal. Letters*, **3**, 43 (1970).
143. S. Lal and G. D. Christian, *Anal. Letters*, **3**, 11 (1970).
144. M. S. Frant and J. W. Ross, *Science*, **167**, 987 (1970).
145. Orion Research Inc., *Newsletter*, **2**, 14 (1970).

133

146. Orion Potassium Electrode Instruction Manual.
147. J. N. Butler and R. Huston, *Anal. Chem.*, **42,** 676 (1970).
148. O. Kedem, M. Furmanski, E. Loebel, S. Gordon and R. Bloch, *Israel J. Chem.*, **7,** 87 (1969).
149. R. M. Carlson and J. L. Paul, *Soil Science*, **108,** 266 (1969).
150. C. J. Coetzee and H. Freiser, *Anal. Chem.*, **40,** 2071 (1968).
151. C. J. Coetzee and H. Freiser, *Anal. Chem.*, **41,** 1128 (1969).
152. E. W. Baumann, *Anal. Chem.*, **40,** 1731 (1968).
153. E. W. Baumann, *Anal. Chem.*, **42,** 110 (1970).
154. Orion Research Inc., *Newsletter*, **2,** 1 (1970).

INDEX

136

137

139

140